Environmental & Biological As

Hesham Yousef (Ed.)
Eman Abdelfattah

Environmental & Biological Assessment of Fertilizer Industry Emissions

Emissions Monitoring Oxidative Stress Biomonitoring Statistical Analysis of Integrated Monitoring and Biomonitoring

Scholars' Press

Cover image: www.ingimage.com

Publisher:
Scholars' Press
is a trademark of
International Book Market Service Ltd., member of OmniScriptum Publishing Group
17 Meldrum Street, Beau Bassin 71504, Mauritius

Printed at: see last page
ISBN: 978-613-8-83811-1

Book title:

Assessment of environmental pollutants emitted through fertilizer industry by using biological technology

Authors

Dr. Eman Alaaeldin Abdelfattah[a] and Prof. Hesham Ahmed Yousef[a]

[a] **Entomology Department, Faculty of Science, Cairo University, Egypt**

Corresponding author:

Dr. Eman Alaaeldin Abdelfattah[a]

Email: eman.alaaeldein@gmail.com

Tele: +201203945934

ACKNOWLEDGEMENTS

First and foremost, I thank Allah, the Most Gracious, the Most Merciful

I wish to thank *Prof. Hesham A. Yousef,* Professor of insect biochemistry and physiology, Entomology Department, Faculty of Science, Cairo University, for his kind assisstance of the present work, and his effective guidance.

I'm grateful to *Prof. Neveen S. Gadallah,* Professor of Taxonomy; and *Mr Yusuf El-demrdash* teacher assistant at Entomology Department, Faculty of Science, Cairo University, for their help in identifying grasshopper samples and their kind cooperation throughout this work.

Great thanks to *Mr. Mustafa Mahmud,* Associate lecturer at Entomology Department, Faculty of Science, Cairo University, for their help in analysis of soil, plant, and water pollutants.

I wish to express my deepest gratitude to *Prof. Tarek M. El-Araby,* Vice Dean of Community Services and Environmental Affairs, Faculty of Science, Cairo University; *Dr. Ahmad F. Youssef,* Head of Cairo University Center for Environmental Hazard Mitigation, and *Chem. AbdEl-hakeem E. El-ghoul* EIMP-Administration and Operation Supervisor Measurements Specialist, for their guidance in analysis of raw data of ambient air quality network and stack emission of Abu-Zaabal Company for Fertilizers and Chemical Industries (Ismalia Agricultural Road).

i

I appreciate *Prof. El-Sayed F. Taha* Dean of Faculty of Science-Cairo University; *Eng. Ahmed Abou Elseoud,* previous Chief Executive Officer of EEAA; *Eng. Mustafa Murad,* Head of Central Administration for Air Quality and Noise Protection; *Dr. Eman A. Zahran,* Director of Ambient Air Department of EEAA; and *Dr. Sherien F. AbdelGhaffar,* Director of Industrial Emission Department, EEAA for their acceptance and permission to get the raw data of ambient air quality network and stack emission of Abu-Zaabal Company for Fertilizers and Chemical Industries (Ismalia Agricultural Road)

Sincere gratitude to *Prof. Abdelhameed El Shaarawi,* Prof. at The American University in Cairo, Egypt (Chair), and *Mr. Hossam M. Hassan,* Associate lecturer at Mathematics Department, Faculty of Science, Cairo University, for their help and instructive guidance in the statistical analysis.

I'm thankful to *Dr. Amir Hassan,* lecturer at Geology Department, Faculty of Science, Cairo University, for their help and instructive guidance in identification Geographic Coordination of the sampling sites using Google earth.

Thanks, are also due to the staff members and personnel of the Entomology Department, especially *Mrs. Amira A. Ammar,* and Faculty of Science, Cairo University staff members, especially *Mrs. Amel M. Abo Zeid* for their help and support.

Finally, but of major importance, deep thanks to *my parents, son, sister, and brother;* they have been my inspiration every day.

CONTENTS

List of Figures

List of Tables

and activity of SOD, CAT, APOX, PPO, POX, GR, and G-S-T antioxidant enzymes) in brain, thoracic muscles, and gut homogenates of males, and females of *Aiolopus thalassinus* collected at different sites located at various distances from fertilizer factory.

instar of *Aiolipus thalassinus*. Samples collected at polluted sites (A-C) located at different distances from fertilizer factory.

ABSTRACT

Phosphate fertilizer industry is considered as one of the main sources of environmental pollutants. Besides solid waste products, e.g. phosphates, sulphates, and heavy metals, also atmospheric pollutants, such as sulphur dioxide (SO_2), nitrogen oxides (NO_2), particulate matter with diameter up to 10 μm (PM_{10}), and hydrofluoric acid fumes (HF) are a serious problem that influences the structure and functioning of ecosystems. The levels of Cd, Pb, Zn, Cu, sulphates and phosphates were measured in soil, plant, and water samples from four sites: a control, that was established at 32 km from the source of pollution and polluted sites that were $\leq$ 1, 3 and 6 km (sites A–C) away from the Abu-Zaabal Fertilizer Company. The highest concentrations of Cu and Zn were found in the soil, plant, and water from site A.

Monitoring of possible impact of fertilizer industry pollutants was assessed using parameters such as hydrogen peroxide concentration (H_2O_2), superoxide anion production rate (O_2^-), DNA strand breaks, protein carbonyls amount, lipid peroxides concentration, levels of non-enzymatic, and antioxidant enzymatic system. The potential use of comet assay, protein carbonylation, and lipid peroxidation as a biomonitoring method of environmental pollution was proposed. The concentration of H_2O_2, and production rate of O_2^- in the brain, thoracic muscles and gut of the males, females, and 5th instar of *Aiolopus thalassinus* from sites A are significantly increased than control site. Strong negative correlation between percentage of cells with visible DNA damage (% of severed cells) and the distance of the sites from fertilizer factory was found. The level of protein carbonyls in the brain, thoracic muscles and gut of the males, females and 5th instar from sites A were 11.82, 4.38, 5.97 (in males), 19.04, 16.65, 7.79 (in females), and 13.31, 18.45, 1.62 (in nymphs) times higher, respectively, compared to the individuals from the control site. Lipid peroxides levels in both sexes and nymph were significantly correlated with the distance from the source of the contamination. In the brain, thoracic muscles and gut of the males and females collected from site A, the level of lipid peroxides was 15.41, 23.49, 11.50 (in males), 25.36, 11.34, 15.37 (in females), and 25.66, 11.45, 15.42 (in nymphs) times higher compared to the values of the control insects. The levels of non-enzymatic such as GSH and activities of antioxidant enzymes such as SOD, CAT, APOX, PPO, POX, GR, and G-S-T were significantly affected by the environmental pollutants. Specific pollution resulting from the activity of the fertilizer industry caused comparable adverse effects in the organisms inhabiting areas up to 6 km from the source of contamination. Therefore, the success of using biochemical parameters as a biomonitoring tools of environmental pollutants levels was discussed.

Keywords: Ecotoxicology; Environmental pollution; Fertilizer company; Oxidative stress; Genotoxicity; Protein carbonyls; Lipid peroxides; Antioxidant enzymes; *Aiolopus thalassinus*; Biomonitors.

1 Introduction

Egypt consumes approximately 14.3 million tons of nitrogenous and phosphate fertilizer per year. Due to its extensive fertilizer production capabilities, Egypt is largely self-sufficient in fertilizer industry and has future investment projects for phosphate fertilizers industry. This due to abundant of phosphate rocks in Egypt and the demand for phosphate fertilizers (IDA, 2014).

The application of fertilizer is associated with higher crop yields and has beneficial effects on the soil and even plants as fertilizers, which usually contain more nutrient that is essential for plant growth and soil fertility (EEA, 2015). Despite the importance of fertilizer application, this industry is considered as one of the main sources of environmental pollutants. Environmental pollution is considered as a serious problem, both in developed and developing countries, so environmental pollution has become an issue of serious international concern (Vij, 2015).

One of phosphate fertilizer production plants in Egypt is the Abu-Zaabal Company for Fertilizers and Chemical Industries (AZFC) where it is location in El Maahd road - Abu Zaabal - Qulubia Governorate. AZFC was established in 1947. They produce concentrated Sulfuric Acid (H_2SO_4) 98% as raw material to produce Phosphate. AZFC consumes raw materials such as Phosphate Ore (24-30% P_2O_5), and Sulfur which their usage is 628,000 and 100,000 m^3/y, respectively. AZFC utilities for its production process water from Ismailia Canal, Natural gas fuel, Self-Generated, and Grid Electricity which it's utilization estimated as 4,500,000 m^3/y, 15,000,000 m^3/y, 25,000, and 25,000 MWh/y, respectively. AZFC produces large

1

amounts of powdered and granulated single super phosphate (17% P_2O_5), granulated triple super phosphate (48% P_2O_5), sulfuric acid (H_2SO_4) and phosphoric acid (48–50% P_2O_5). They were produced in 585,000, 87,000, 300,000 and 41,000 tons per year, respectively (Report of PMU, 2012).

The main pollutants from phosphate fertilizer industries are: hydrofluoric acid (HF), sulphur dioxide (SO_2) (29 mg/m^3, comparing to the Egyptian maximum limits; 60 mg/m^3 in the air), nitrogen dioxide (NO_2) (32 mg/m^3, comparing to the Egyptian maximum and particulate matter of 10 mm diameter (PM_{10}) (ESE, 2014). Also, dust particles emitted from phosphate fertilizer industries form a complex of organic compounds and minerals. Transport and deposition of such pollutants may have a hazardous effect on the environment, particularly air, soil, plant, and water. Phosphorus species are the principal carriers of heavy metals such as Zn, and Cd in soils, therefore the phosphate industry is considered as one of the soil pollution hazards. These contaminants can be potentially hazardous to terrestrial organisms and groundwater (Kassir et al., 2012).

Environmental stress primarily increases the production of reactive oxygen species (ROS) in organisms. In aerobic cells, ROS are produced from molecular oxygen as result of normal cellular metabolism (Hermes-Lima, 2004; Sena and Chandel, 2012). Exogenous sources, such as heavy metal emitters, chemicals and drug manufacturers, domestic sewage, polymer and petrochemical-based industries, oil refineries, mining, fertilizer industries and many others can, directly and indirectly, significantly influence the level of ROS in cells (Amado et al., 2006; Sureda et al., 2006; Dos Anjos et al., 2011). The three major ROS that are of physiological significance are superoxide anion ($O_2^{\cdot-}$), hydroxyl radical ($^{\cdot}OH$) and hydrogen peroxide (H_2O_2) (Equation 1-5) (Gilbert 2000). Through evolution and natural selection, organisms have developed efficient antioxidant mechanisms, which allow them to control their ROS levels, thus limiting the damage to cells. The levels of antioxidants in organisms can increase to reverse the imbalance that is caused by pro-oxidant substances and these levels can serve as biomarkers of oxidative stress (Livingstone, 2001). However, cellular antioxidant defense systems can be impaired when organisms are exposed to severe or prolonged environmental stress.

$$O_2 + e^- \rightarrow O_2^{\cdot-} \qquad\qquad\qquad 1$$

$$O_2^{\cdot-} + H^+ \rightarrow HO_2^{\cdot} \qquad\qquad\qquad 2$$

$$HO_2^{\cdot} + e^- \rightarrow H^+ + H_2O_2 \qquad\qquad 3$$

$$H_2O_2 + e^- \rightarrow OH^{\cdot} + OH^- \qquad\qquad 4$$

$$\cdot\,OH + e^- \xrightarrow{H^+} H_2O \qquad\qquad\qquad 5$$

When antioxidant defenses are impaired or overloaded, oxidative stress cause serious damage to macromolecules in living organisms, including DNA damage, enzyme inactivation, protein carbonylation, and the peroxidation of cell components, especially lipids (Halliwell and Gutteridge, 1984) (Equation 6-9). Protein carbonyls amount, and lipid peroxide concentration considered as biomarker of oxidative stress (Gutteridge, 1995; Dalle-Donne et al., 2003; Hermes-Lima, 2004; Migula et al., 2004; Birben et al., 2012; Kaviraj et al., 2014).

3

$$LH + OH^{\cdot} \rightarrow L^{\cdot} + H_2O \quad \textbf{(Initiation)} \qquad\qquad 6$$

$$L^{\cdot} + O_2 \rightarrow LOO^{\cdot} \quad \textbf{(Propagation)} \qquad\qquad 7$$

$$LOO^{\cdot} + LH \rightarrow L^{\cdot} + LOOH \quad \textbf{(Propagation)} \qquad\qquad 8$$

$$LOO^{\cdot} + LOO^{\cdot} \rightarrow LOOL + O_2 \quad \textbf{(Termination)} \qquad\qquad 9$$

Total genotoxic potential in the environment can be measured using single cell gel electrophoresis (SCGE), comet assay. This method is considered as one of the simplest, most sensitive and reliable method for detecting DNA strand breakages. High sensitivity of the comet assay allows early detection of the deleterious effect of pollutants (Jha, 2008; Al-Shami et al., 2012; Guanggang et al., 2013). DNA damage can be measured in the cells of organisms (Rojas et al., 1999; Dhawan et al., 2009). For DNA oxidative damage, the values of tail length (TL), tail moment (TM), and % DNA in tail (TDNA) are the most informative (Tice et al., 2000; Lovell and Omori, 2008; Carmona et al., 2011; Augustyniak et al., 2016a). In the assessment of the oxidative DNA damage, the percentage of severed cells may be considered as a useful supplementary parameter (Bilbao et al., 2002; Augustyniak et al., 2016b). DNA damage may be considered as a signal of disturbances occurring at the molecular level that may cause chromosomal instability and subsequent pathological changes in the cells and tissues, e.g. morphological abnormalities, cancer diseases, reduction in gamete production, and finally, population extinction (Jha, 2008; Dhawan et al., 2009). The comet assay applications involve genotoxicity studies, biomonitoring, ecotoxicology, as well as basic research on DNA damage and its repair (Rojas et al., 1999; Jha, 2008; Dhawan et al., 2009; Collins et al., 2014). Recently, the comet assay has become more common as a tool to study genotoxicity of environmental pollutants in different animals, and in the last decade, also in insects (Mukhopadhyay et al., 2004; Siddique et al., 2005; Yousef et al., 2010; Carmona et al., 2011; Sharma et al., 2011; Shukla et al., 2011; Guanggang et al., 2013; Abdelfattah et al., 2017).

Protection against xenobiotics, including ROS-mediated environmental pollutants, can be realized by two main mechanisms: (i) the avoidance of stress, which cannot be achieved by organisms living in polluted areas or (ii) the intensification of the antioxidative defense of the organism (Migula et al., 2004). Antioxidants response - which includes non-enzymatic such as glutathione reduced (GSH), α-tocopherol, and β-carotene, and enzymatic antioxidants such as superoxide dismutase (SOD), catalase (CAT), glutathione peroxidase (GPx), ascorbate peroxidase (APOX), polyphenoloxidase (PPO), glutathione reductase (GR), acetylcholine esterase (AChE), and glutathione-s-transferase (GST) - are supposed to be important indicators of oxidative stress (Livingstone, 2001; Lushchak, 2011) (Equation 10-18). Moreover, acetylcholinesterase (AChE), a key enzyme in the nervous system used as a biomarker of exposure to organophosphate and carbamate pesticides, was affected in *Aiolopus thalassinus* treated with metals (Schmidt and Ibrahim, 1994). A recent study indicated that the AChE activity measured in grasshoppers can used as a biomarker of arsenic pollution (Nath et al., 2015).

$$2O_2^{\cdot-} + 2H^+ \xrightarrow{\text{SOD}} H_2O_2 + O_2 \qquad\qquad \mathbf{10}$$

$$2H_2O_2 \xrightarrow{\text{CAT}} H_2O + O_2 \qquad\qquad \mathbf{11}$$

$$2H_2O_2 \xrightarrow{\text{APOX}} H_2O + O_2 \qquad\qquad \mathbf{12}$$

5

$$\text{Tyrosine} + \text{D}OPA + O_2 \xrightarrow{\text{PPO}} \text{DOPA} + \text{Dopa quinone} \qquad \textbf{13}$$

$$AH_2 + H_2O_2 \xrightarrow{\text{POX}} A + 2H_2O \qquad \textbf{14}$$

$$H_2O_2 + 2GSH \xrightarrow{\text{GPx}} 2H_2O + GSSG \qquad \textbf{15}$$

$$LOOH + 2GSH \xrightarrow{\text{GPx}} LOH + H_2O + GSSG \qquad \textbf{16}$$

$$GSSG + NADPH + H^+ \xrightarrow{\text{GR}} 2GSH + 2NADP^+ \qquad \textbf{17}$$

$$LOOH + 2GSH \xrightarrow{\text{GST}} LOH + H_2O + GSSG \qquad \textbf{18}$$

Insects common in terrestrial ecosystems, such as grasshoppers, can be seen, as sensitive to environmental changes. They can be considered as an interesting subject of ecotoxicological research, and a biomonitor of environmental pollutants, including heavy metals, near an industrial region (Chen et al., 2005; Azam et al., 2015). Moreover, grasshoppers are widespread in strongly industrialized areas (Augustyniak and Migula, 1996, 2000). One example of grasshopper is *Aiolopus thalassinus* that used in ecotoxicological research, mainly to monitor the impact of heavy metals on the fertility and fecundity of grasshoppers, as well as the growth and development of the progeny (Schmidt et al., 1992).

In the present study, we want to answer the question if continuous contact of *Aiolopus thalassinus* individuals with specific pollutants, resulting from fertilizer industry activity, influences production of reactive oxygen species, stability of macromolecules (such as DNA, protein, and lipid), activation of non-enzymatic and enzymatic antioxidants in different tissues of the insect.

The aim of the present work is: (i) Determination of environmental pollutants concentration around the area of AZFC, including: air pollutants (PM_{10}, and SO_2) in ambient air and (TSP, and SO_2) stack emission of AZFC), Soil, plant, and water pollutants (heavy metals such as Cu, Zn, Pb, and Cd), PO_4^{3-}, and SO_4^{2-}, (ii) Evaluation H_2O_2 concentration, O_2^{-} production rate, DNA damage, protein carbonyls level, lipid peroxides concentration, (iii) Evaluation the cellular response to oxidative stress as non-enzymatic antioxidants (GSH) concentration, and activity of antioxidant enzymes (SOD, CAT, APOX, PPO, POX, GR, and G-S-T) in brain, thoracic muscles, and gut of male, female, and 5th instar of *Aiolopus thalassinus* individuals inhabiting sites at different distances from the fertilizer industry and comparing to insects from control site. Consequently, the success of using these biochemical parameters as biomonitoring for environmental pollution level.

2 Materials and Methods

2.1 Study area

Insects were collected from four sites located at various distances from Abu-Zaabal Company for Fertilizers and Chemical Industries – the main source of contamination. Specific conditions in the area allowed to set experimental plot along a pollution gradient. Three polluted sites (A-C) of the cultivated spots of this area were located nearly ≤ 1, 3, and 6 km away from Abu-Zaabal Company, respectively (along a branch of the Nile river). Control site was established about 32 km from the source of pollution – at the Cairo University Campus (Fig. 1). Geographical location and the dominant vegetation type of sampling sites are shown in Table 1.

Table 1. Geographical location and the dominant vegetation type of sampling sites (distance from Abu-Zaabal Company in brackets).

Sites	Control site	Site A (1 Km)	Site B (3 km)	Site C (6 km)
Latitude	30°1′48.76″N	30°16′3.98″N	30°16′33.44″N	30°18′4.16″N
Longitude	31°11′23.07″E	31°22′6.68″E	31°23′4.83″E	31°23′39.62″E
Dominant vegetation type	Alfalfa, *Medicago sativa* (Papillioacea)			

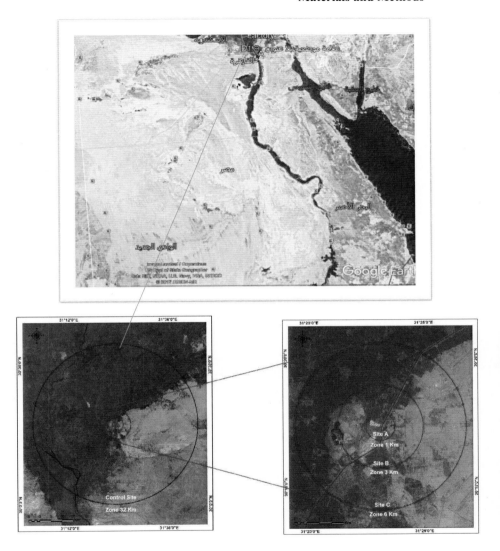

Fig. 1. Map showing study area location.

Lined polygon represents the fertilizer factory, red spots represent location of sampling sites

9

2.2 Determination of pollutants concentrations in studied areas

2.2.a Air pollutants concentration

Data of air pollutants were obtained from Egyptian Environmental Affairs Agency, Ministry of Environment, Arab Republic of Egypt. The data include ambient air quality network (Sulphur dioxide (SO_2) and particulate matter with diameter 10 µM (PM10) and stack emission (SO_2 & total suspended particulate (PM10) of Abu-Zaabal Company for Fertilizers and Chemical Industries.

2.2.b Soil, plant, and water pollutants concentration

2.2.b.I Soil, plant, and water sampling

Surface soil samples (0–20 cm depth) were collected from each location (control sites and polluted sites (Site A- C) using a clean stainless-steel spoon. Approximately 10 kg of soil per site were collected from various points in a zig-zag pattern over the area. The arbitrary selection of sample locations requires each sample location to be chosen independently, so that results in all locations within the area of concern have an equal chance of being selected. To facilitate statistical probabilities of contaminant concentration, the area of concern must be homogeneous with respect to the parameters being monitored (Mason, 1983). Then from five to seven point samples were mixed together in plastic bags. This procedure was repeated three times and finally, three mixed soil samples were analyzed. After preliminary debris removal, the soil samples were dried and sieved using a mechanical sieve.

For plant samples, we used the same grid locations as it will provide the potential for comparison of contaminant concentrations in the soil and the vegetation (Bonham, 1989).

Surface water samples were collected at each sample sites location using direct dipping stainless steel scoops which provide a means of collecting surface water samples from surface water bodies that are too deep to access by wading. Stainless steel scoops are useful for reaching out into a body of water to collect a surface water sample. The scoop may be used directly to collect and transfer a surface water sample to the sample container, or it may be attached to an extension to access the selected sampling location (US-EPA, 2013).

2.2.b.II Determination of heavy metals, SO_4^{2-} and PO_4^{3-} concentration in soil and plant digested samples, and water samples

Soil, and plant samples were digested according to Moor et al., (2001). Briefly, 0.5 g of samples pre-digested with 12 mL 37% HCl: 65% HNO_3 (3:1) mixture for 24 h at room temperature. Then, mixture solution of 2.5 mL of 37% HCl and 2.5 mL of 30% H_2O_2 were added to complete the digestion. Finally, the suspension was filtered and stored at 4°C for determination of heavy metal concentration.

Heavy metal content was analyzed according to Boon and Soltanpour (1991), using inductively coupled spectrometry plasma atomic emission spectroscopy (ICP-AES; Model Ultima 2; Jobin Yvon). Sulphates and phosphates were analyzed in compliance with the method of Radojevic and Bashkin (1999). Absorbance was measured at 420 nm using a UV/Vis Jenway- 7305 spectrophotometer

(Bibby Scientific Limited, Staffordshire, UK). The concentrations of the heavy metals, SO_4^{2-} and PO_4^{3-} in the soil, plant, and water samples were measured in three replicates.

3. Insects samples

Preliminary insect sampling was performed to check the quantitative structure of the grasshopper population so as proper material for further studies would be accessible. Adult (males and females) and 5th instar of grasshoppers, *Aiolopus thalassinus* were collected with a sweep-net from four sites with different levels of pollution. Insects were transported to the laboratory in small 30 cm × 30 cm × 30 cm cages (approximately 25 insects per cage), and stored at −20 °C until use.

4. oxidative stress

4.I Reactive oxygen species assay

4.I.a Hydrogen Peroxide concentration (H_2O_2)

The H_2O_2 concentration of experimental samples was determined spectrophotometrically according to the method of Junglee et al. (2014). 150 mg of each tissue samples was directly homogenized with 1 mL of solution containing 0.25 mL Trichloroacetic acid (TCA) (0.1% (w:v)), 0.5 mL KI (1 M) and 0.25 mL potassium phosphate buffer (10 mM, pH 7.0) at 4°C for 10 min (one-step buffer: extraction and colorimetric reaction combined). At the same time, for every sample, a control was prepared with H_2O instead of KI for tissue coloration background. Aluminum foil was taken to protect samples

12

and solutions from light. The homogenate was centrifuged at 12,000 × g for 15 min at 4°C. 200 μL of supernatant from each tube were placed in UV-microplate wells and left to incubate at 20°C for 20 min. Samples and blanks were analyzed in triplicate. A calibration curve obtained with H_2O_2 standard solutions prepared in 0.1% TCA was used for quantification.

4.I.b Superoxide anion ($O_2^{\cdot-}$) production rate

The $O_2^{\cdot-}$ production rate of experimental samples was determined spectrophotometrically according to the method of Chen and Li (2001). The level of superoxide anion radical was determined by the rate of conversion of epinephrine to adrenochrome with 1 mM NADPH as substrate. The reaction mixture contains 1 mM epinephrine, 1 mM NADPH, and 50 mM potassium phosphate buffer (pH 7.0), and 150 μg protein of tested samples. The absorbance difference (A_{485}- A_{575}) was recorded.

4.II Oxidative damage assays

4.II.a DNA damage assay-Comet assay

Insects were dissected to isolate brain, thoracic muscles, and gut tissues for further determination of DNA damage using comet assay. For each experimental group, three slides of a 5 insects pool of both sexes and 5th instar of *A. thalassinus* were prepared. So, the total number of females, males, and 5th instar collected from each site were 15.

The alkaline Single Cell Gel Electrophoresis assay (SCGE), known as the Comet assay was used to assess the DNA strand breaks according to Yousef et al. (2010). Immediately after dissecting, tissues were macerated in 0.5 mL of 1x PBS with a teasing needle for 30 seconds. The PBS composition was as follows: 0.8 g NaCl, 0.02 g KCl, 0.144 g Na_2HPO_4, 0.024 g KH_2PO_4 diluted in 100 mL of distilled water; pH was adjusted to 7.4 by adding 2 M HCl or NaOH. The macerates were resuspended in 3 mL of 1x PBS and centrifuged (1500 rpm, 4 °C, 5 min), and then pellet was again suspended in 0.5 mL of 1x PBS (Martínez-Paz et al. 2013; Morales et al., 2013). The average number of cells in the suspension was in a range from 100 to 150 cells/mL. Cell suspensions (60 μL) were mixed with 50 μL of 1% low melting-point agarose, 110 μL of the mixture was spread on a microscopic slide previously covered with a layer of 0.8% regular melting-point agarose.

After solidification of the cells-agarose layer at 4°C, the slides were immersed for 24 hours (Tice and Vazquez, 1999) in a fresh lysis solution (164 g NaCl, 37 g of ethylene-diamine tetracetic acid (EDTA), 1 g Tris base merged into 890 mL of distilled water and stirred before adding 8 g of NaOH; pH 10.0; 4 °C. Then 10 mL of TritonX-100 and 100 mL of dimethyl sulfoxide was added before use. After lysis, the slides were washed two times with distilled water, and immersed for 5 minutes at 4 °C in a freshly prepared alkaline electrophoresis buffer (30 mL of 10 N NaOH, 0.5 mL of 200 mM EDTA mixed with 1000 mL of distilled water; pH was adjusted to 13.0 using 2 M HCl or NaOH).

Electrophoresis was performed at 20 V for 20 min. Next, the slides were immersed in neutralization buffer for 15 minutes (Tris base; pH was adjusted to 7.5 using 2 M HCl), drained by immersing in cold absolute ethanol for 5 minutes, and stored under dry conditions. Before the analyses, slides were stained with 40 μL of ethidium bromide solution. (2 μg/mL). An analysis of DNA damage was performed using OPTIKA B-350 fluorescent microscope (OPTIKA, Ponteranica, Italy), with a CCD camera.

The image analysis system (Comet IV) was used to measure DNA damage level. The percentage of DNA in the comet tail (TDNA; defined as the total comet tail intensity divided by the total comet intensity, and multiplied by 100), the length of the comet tail (TL; described as the comet head diameter subtracted from the overall comet length), as well as percentage of severed cells (% severed cells; the number of cells with DNA damage) were recorded. Olive Moment (OM; defined as product of tail DNA% and the distance between the intensity-weighted centroids of head and tail) and Tail Moment (TM; defined as Tail length times Tail DNA%) were also estimated using Comet Score software and included into the statistical analysis (Gyori et al. 2014; Comet Score Tutorial). For each site, insect samples, and tissue 3 slides and 50 cells per slide were analyzed. A total number of 108 slides were analyzed.

4.II.b Protein carbonyls and lipid peroxides assays

The brain, thoracic muscles and gut tissues were isolated and homogenized in 5 mL of an ice-cold phosphate buffer with additives (60 mL of 50 mM phosphate buffer, 10 mL of 0.1% TritonX-100, 5 mL of 0.05 mM $CaCl_2$; after adjusting pH to 7.0 with HCl or NaOH, the mixture was filled with distilled water to a volume of 100 mL). After homogenization (mortar, 10 strokes/30s), the samples were centrifuged at 2000×g for 10 min at 4 °C. Then, 800 μL aliquots of the supernatant were transferred to a clean microtube with 800 μL of 30% trichloroacetic acid (TCA). The samples were incubated for 30 min at room temperature and then centrifuged at 5000×g for 10 min at 4 °C. Protein carbonyls assay was conducted on precipitated pellets, while the assay of lipid peroxides was conducted on the supernatant. The concentrations of lipid peroxides and protein carbonyls were determined using the method of Hermes-Lima et al. (1995) and Levine et al. (1990), respectively. The measurements were done in three replicates (each replicate was a pool of 10 insects).

In protein carbonyls assay, 800 μL supernatant aliquot of each tissue extract, was transferred to a clean microtube with 200 μL of 10 mM 2, 4-Dinitrophenylhydrazine (DNPH) prepared in 2 M HCl. The samples were incubated for 30 minutes at room temperature, precipitated with 10% TCA, and left for 10 min at room temperature. The samples were centrifuged at 5000 ×g for 7 min at 4 °C. The precipitated pellet was washed four times with an ethanol/ethyl acetate (1:1) mixture, and redissolved in 3.5 mL of sodium phosphate buffer (60 mL of 150 mM sodium phosphate buffer, 30 mL of 3% sodium dodecyl sulphate, adjusted to a final volume of 100 mL with distilled water after adjusting the pH to 6.8 with 2M HCl). Insoluble material was removed by centrifugation at 2000 ×g for 1 min. Finally, the absorbance was measured at 366 nm, and the rate of protein carbonyls concentration was expressed as OD/μg protein. Blank was similarly prepared and treated except for DNPH that was not added to the samples.

For lipid peroxides, 1 mL supernatant aliquot was used for the assay. The following components were sequentially added to the samples: 1750 µL of 1 mM FeSO$_4$, 700 µL of 0.25 M H$_2$SO$_4$, and 700 µL of 1 mM xylenol orange. Samples were then incubated under dark conditions at room temperature for 3 h. The initial absorbance was measured at 580 nm. Then, 10 µL of 0.5 mM cumene hydroperoxides (as an internal standard) was added to each sample, and the samples were maintained at room temperature for 1 h before the absorbance was re-measured at 580 nm. The change in absorbance due to addition of internal standard was calculated. Lipid peroxides concentration was expressed as mM cumene hydroperoxides/µg protein.

5. oxidative stress response

5.I Non- enzymatic antioxidants assay

5.I.a Glutathione reduced (GSH)

The procedure of Allen et al. (1984) was adapted for GSH determination. In briefly, 200 mg of each tissue samples were homogenized with 2 mL of 5% (w/v) TCA in 1 mM EDTA and then centrifuged at 10,000 ×g for 20 minutes at 4°C. 1 mL of the reaction mixture, containing 150 µL extract, 800 µL of 0.1 M phosphate buffer (pH 8.0), and 50 µL of DTNB (0.01% in 0.1 M phosphate buffer, pH 8.0), was mixed thoroughly and then incubated at 25°C for 20 minutes. The absorbance of the reaction mixture was 412 nm. The GSH content was determined from a GSH standard curve, and the result was expressed in µg GSH/ mg protein.

17

5.II Enzymatic antioxidants assays

After dissection, the brain, thoracic muscles and gut tissues were homogenized (mortar, 10 strokes/30 s) in an ice-cold phosphate buffer (w/v ratio 1:4), the pH of the buffer was adjusted to pH= 7.0 using 2 M NaOH or 2 M HCl). The homogenates were centrifuged at 10,000×g for 30 min at 4 °C and the supernatants were used to measure the activities of the antioxidant enzymes.

5.II.a Superoxide dismutase (SOD)

SOD activity was measured based on the procedure described by Misra and Fridovich (1972). The reaction mixture was as follows: 402 µL of a sodium carbonate buffer (200 mM; pH 10.0), 35 µL of EDTA (10 mM), 87 µL of the supernatant of the appropriate tissue and 2.8 mL of freshly prepared epinephrine (15 mM). The absorbance was measured at 480 nm using a UV/Vis Jenway-7305 spectrophotometer (Bibby Scientific Limited, Staffordshire, UK). SOD activity was expressed as OD/µg protein/min.

5.II.b Catalase (CAT)

The activity of catalase (CAT) was assessed in compliance with the method of Aebi (1984). The reaction mixture contained 3060 µL of a potassium phosphate buffer (50 mM, pH 7.0), 510 µL of the supernatant and 40 µL of freshly prepared H_2O_2 (10 mM). The decrease in absorbance was measured at 240 nm and the CAT activity was expressed as the decrease of OD/µg protein/min.

5.II.c Ascorbate peroxidase (APOX)

The activity of ascorbate peroxidase (APOX) was determined according to Asada (1984). The reaction mixture consisted 3 mL of phosphate buffer (100 mM, pH 7.0) containing 0.1 mM EDTA, 0.3 mM ascorbic acid, 0.06 mM H_2O_2 and 0.1 mL of the enzyme extract from each sample tissue. Distilled water instead of enzyme extract served as the controls. The change in absorbance was recorded using a spectrophotometer at 290 nm over a period of 30 s after the addition of H_2O_2.

5.II.d Polyphenoloxidase (PPO)

The method of Kumar and Khan (1982) was used to estimate the polyphenoloxidase (PPO) activity. In a reaction mixture containing 2 mL of a potassium phosphate buffer (0.1 M, pH 6.0), 1 mL of 0.1 M catechol and 0.5 mL of the enzyme extract. The purpurogallin that was formed was read at 495 nm. PPO activity was expressed as OD/min/mg protein.

5.II.e Peroxidase (POX)

The peroxidase activity was estimated according to the method described by Mazhoudi et al., (1997)   with minor modifications. The reaction mixture contained 50 mM potassium phosphate buffer (pH 7.0), 1% (m/v) guaiacol, 0.4% (v/v) H_2O_2, and 0.1 mL of the enzyme extract from each sample tissue. Changes in the absorbance were measured at 470 nm. POX activity was expressed as OD/min/mg protein.

5.II.f Glutathione reductase (GR)

The activity of GR was determined according to Carlberg and Mannervik (1985) with the following minor modifications. The reaction mixture contained 1.7 mL of 2 mM oxidized glutathione (GSSG), 175 µL potassium phosphate buffer (50 mM, pH adjusted at 7.5 with 2 M HCl or NaOH), 875 µL of 3 mM DTNB, 175 µL of 2 mM NADPH, and 350 µL supernatant of the appropriate tissue. The absorbance was measured at 420 nm, and the GR activity was expressed as OD/µg protein/min.

5.II.g Glutathione-S-Transferase (G-S-T)

The activity of GST was determined according to method of Seyyedi et al. (2005) with minor modification. Reduced glutathione (GSH) and 1- chloro-2, 4-dinitrobenzene (CDNB) were used as substrates. The reaction mixture contained 100 µL of supernatant of the appropriate tissue sample, and 900 µL of the following mixture (882 µL PBS pH 7.0, 9 µL of 100 mM CDNB and 9 µL of 100 mM GSH). The absorbance was measured at 340 nm. G-S-T activity was expressed in OD/min/mg protein.

The total protein concentration of samples was determined spectrophotometrically according to the method of Bradford (1976), with Coomassie Brilliant Blue (COBB). 5 ml of the dye reagent (10 mg COBB + 5 ml methanol + 10 ml 85% O-phosphoric acid, completed to 100 ml with distilled water) were added to 100 µl of each sample in a separate test tube. The contents of the tube were mixed by gentle shaking and left to stand for 2 min. The OD of the protein sample was measured at 595 nm against a blank of a tube containing distilled water instead of the protein sample. Bovine serum albumin (BSA) fraction V (Sigma-Aldrich) dissolved in 0.15 M NaCl was used as protein standard.

6. Statistical analysis

Wald-wolfowitz run test was performed to examine randomness of data obtained from experiments. Also, Levene's test of equality of error variances, and by the Kolmogorov–Smirnov test were used to examine homogeneity of variance, and data normality respectively.

Non-parametric tests were carried out using the Mann–Whitney test for two median values and the k independent Kruskal–Wallis test for more than two median values. These non-parametric tests were used to air pollutants in ambient air (SO_2 and PM_{10}), stack emission (SO_2 and TSP) from fertilizer factory, macromolecule damage (DNA damage parameters, protein carbonyls, and lipid peroxides), and antioxidant enzyme assays are expressed using median and quartile deviation (25th and 75th percentiles: P25 and P75).

Time series statistical analysis were used in data of ambient air quality network and stack emission of AZFC to examine stationary of the data. The plotting of autocorrelation factor (ACF) and partial autocorrelation factor (PACF) help to determine the order of AR and MA terms respectively. ACF plot is useful in determining the type of model to fit to a time series of length N. PACF is used to measure between an observation k period ago and the current observation. The process of establishing ARIMA model includes analysis and giving supposed model, estimating the model parameters, diagnosing and testing the model, and finally confirmed the model.

A parametric test was performed using a one-way analysis of variance (ANOVA, *Tukey's-b* test, $p < 0.05$). This parametric test was applied on the data of heavy metal, sulphate, and phosphate concentrations in soil, plant, and water samples which were presented as the mean $\pm$ SE.

Generalized Estimating Equation (GEE) was used to examine the effect of distance from the fertilizer company, types of tissues, sex and the interactions of these variables on reactive oxygen species (ROS) (H_2O_2 concentration, and O_2^- production rate), macromolecule damage (DNA strand breaks; comet parameters (TL, TDNA, OM, TM and % of severed cells), Protein carbonyls and lipid peroxides), and oxidative stress response (non-enzymatic assay (glutathione reductase concentration), and activity of antioxidant enzymes (SOD, CAT, APOX, PPO, POX, GR, G-S-T).

Correlations between the distance from the Abu-Zaabal Fertilizer Company and the biochemical results (reactive oxygen species, DNA strand breaks, protein carbonyls, lipid peroxides, concentration of non-enzymatic antioxidants the activity of antioxidant enzymes) were performed based on Pearson's regression analysis using multiple regression models.

Hierarchical Cluster Analysis (HACA) based on agglomerative statistics using Ward's Method was calculated for reactive oxygen species, DNA strand breaks, protein carbonyls, lipid peroxides, concentration of non-enzymatic antioxidants the activity of antioxidant enzymes. The goal of HACA is finding possible clusters or groups

22

among the observational units, based on level of similarities and differentiations (Azam et al., 2015). At each stage, the average similarity of the cluster is measured. The difference between each case within a cluster and that average similarity is calculated.

Cross-tabulation is one of the most useful analytical tools which use Chi-square statistic. It is most often used to analyse categorical (nominal measurement scale) data. Chi-square tests whether the two variables are independent. The variables are independent (have no relation), when accept the null hypothesis ($p > 0.05$). Chi-square (X^2) test using cross tabulation to test independency (i) between reactive oxygen species amount (ROS) (production rate of superoxide anion (O_2^-) and concentration of hydrogen peroxide (H_2O_2)) and macromolecules damage (DNA strand breaks, protein carbonyls amount, and lipid peroxides concentration), (ii) air, soil, plant, and water pollutants concentration (heavy metals (Cu, Zn, Pb, and Cd), and PO_4^{3-}, and SO_4^{2-}) and macromolecules damage (DNA strand breaks, protein carbonyls amount, and lipid peroxides concentration) (iii) independency of each pollutants on the other.

All statistical analyses were performed using IBM SPSS Statistics for Windows (Version 17.0. Armonk, NY: IBM Corp.).

3 Results

3.1 Determination of pollutants concentration in studied areas

3.1.a Air pollutants concentration

The ambient air pollutants, SO_2, and PM_{10} (particulate matter with diameter 10 μm) in Abu-Zaabal Company for Fertilizers and Chemical Industries (AZFC), were measured using ambient air quality network which located at Abu-Baker El-Sedeek School (30°15'14.5" N; 31°21'6.6" E), approximately 1.5 km away from AZFC. It was revealed a significant difference among pollutant concentrations (monthly average) (k independent Kruskal–Wallis test, $P<0.05$), except for SO_2 among August, September, October, and November 2015, among May, June, July, and August in 2016, and for PM_{10} among April, and November in 2016. The mean daily concentration of SO_2 never exceed 150 $\mu g/m^3$ at 24 h., but it exceeds for PM_{10} (Table 2).

The main pollutants in AZFC stack emission are SO_2, and total suspended particles (TSP). There was no significant difference among these pollutants concentration (monthly average) (k independent Kruskal–Wallis test, $P>0.05$) on September, and November 2015, among (January, February, and March); (May, and July); (October, and November) in 2016, and for TSP among all examined months in 2015, and 2016 except August in 2016 (Table 3).

The time series modelling of average concentrations of ambient air in AZFC (Fig. 2 and 5). showed that the series had not have seasonal fluctuations with somewhat a constant trend. The continuous line (sampling) indicates time periods when SO_2, and PM_{10}

24

samples were collected with the sequential samplers at the respective sites. The missing parts in line (no sampling) show phases during which no filter sampling was carried out. To determine a proper model for a given time series data. It is necessary to carry out the autocorrelation factor (ACF) and partial autocorrelation factor (PACF) analysis. Fig. (3 and 6) showed that there are significantly large sample ACF values at increasing lags, which do not diminish quickly. This indicates that the non-stationary of the data, which proved the stationarity of data as ACF decrease to zero quickly. Based on the data from Fig. (4 and 7), we can use ARIMA model as the supposed model to predict the data in Fig. (2 and 5), is reasonable. The final step, diagnosis and testing a supposed model through analysing residual series between forecasted series and original series, it can be concluded that the residual series are appeared as a white noise series as ACF, and ACF in Fig. (4 and 7) are nearly equal to zero. This implies that the supposed ARIMA model is valid for data in Fig. (2 and 5).

The results of ambient air quality network pollutants and stack emission in AZFC (SO_2, and PM_{10} for ambient air; SO_2, and TSP for stack emission) showed that the significant level of $\varkappa^2$ among these pollutants were lower than 0.05; and person's R among these pollutants were -0.12, and -0.22 in ambient air, and stack emission, respectively. This means that there is a weak negative correlation between SO_2, and PM_{10} in ambient air, and between SO_2, and TSP in stack emission.

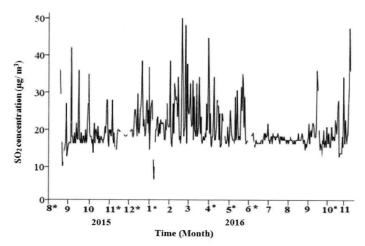

Fig. 2: Time series of SO₂ mass concentrations at Abu-Zaabal ambient
air quality network monthly averages of daily monitoring data.
* means that data not covered all the days.

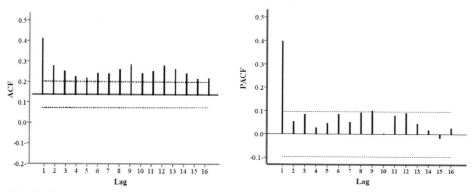

Fig. 3: The autocorrelation function plot (ACF), and partial autocorrelation function
(PACF) of SO₂ mass concentrations at Abu-Zaabal ambient air quality
network monthly averages of daily monitoring data. The results show that
how correlation in the time series varies with the distance between time
points. Dash lines represent the upper and lower confidence limit coefficient.

26

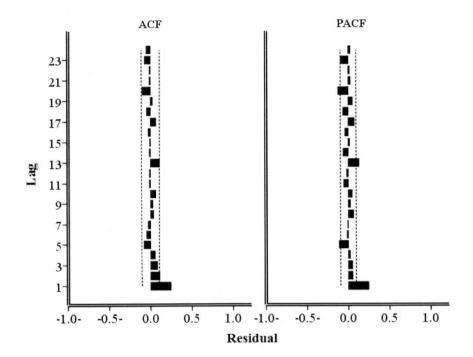

Fig. 4: The autocorrelation function (ACF), and partial autocorrelation function plot (PACF) of the residual series between the forecasted series and the real (differential) series of SO₂ mass concentrations at Abu-Zaabal ambient air quality network monthly averages of daily monitoring data. Dash lines represent the upper and lower confidence limit coefficient.

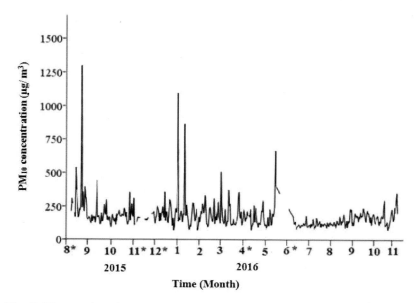

Fig. 5: Time series of PM_{10} mass concentrations at Abu-Zaabal ambient air quality network, monthly averages of daily monitoring data.
* means that data don't cover the whole days.

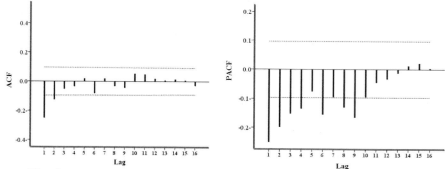

Fig. 6: The autocorrelation function plot (ACF), and partial autocorrelation function (PACF) of PM_{10} mass concentrations at Abu-Zaabal ambient air quality network monthly averages of daily monitoring data. The results Show that how correlation in the time series varies with the distance between time points.
Dash lines represent the upper and lower confidence limit coefficient.

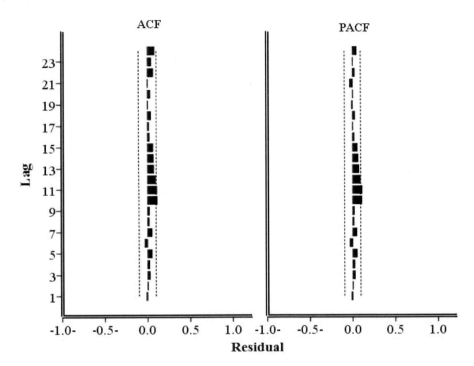

Fig. 7: The autocorrelation function (ACF), and partial autocorrelation function plot (PACF) of the residual series between the forecasted series and the real (differential) series of PM_{10} mass concentrations at Abu-Zaabal ambient air quality network monthly averages of daily monitoring data. Dash lines represent the upper and lower confidence limit coefficient.

Table 2: Median, percentile deviation (P25 and P75), maximum, and minimum ($\mu g/m^3$) of concentration of air pollutants (SO_2 and PM_{10}) in ambient air quality network of fertilizer factory from August 2015 to November 2016.

Year		2015					2016										
Pollutants	Parameter	Aug.	Sep.	Oct.	Nov.	Dec.	Jan.	Feb.	Mar.	Apr.	May	Jun.	Jul.	Aug.	Sep.	Oct.	Nov.
SO_2	Median	14a	15ab	16b	16b	18	20	24	24	17b	16ab	16ab	16ab	15a	15a	15a	16ab
	P25	12	14	14	13	16	17	17	17	14	15	14	15	15	15	13	14
	P75	16	17	17	17	22	21	30	30	23	20	25	16	15	17	16	22
	Min.	7	14	14	12	16	15	14	14	13	13	13	14	14	12	10	10
	Max.	43	43	43	18	29	26	52	52	46	30	25	20	17	20	36	49
PM_{10}	Median	215	157	165	144	185	178	169	139	147*	119	144	89	104	130	161	148*
	P25	153	129	134	122	142	126	120	107	109	98	106	81	93	101	141	112
	P75	308	174	185	172	222	238	217	231	178	187	360	99	114	166	186	203
	Min.	112	75	96	97	87	53	54	81	57	74	97	69	71	71	96	61
	Max.	1323	437	349	300	351	1114	325	503	352	284	667	168	152	221	229	345

In the same raw: the same letter, and * (for SO_2 and PM_{10} respectively) indicates no significantly different among month (k independent Kruskal–Wallis, $P>0.05$) in each year separately. R^2 was 0.245, and 0.023 for PM_{10}, and SO_2, respectively. Raw data obtained from EEAA.

Table 3: Median, percentile deviation (P25 and P75), maximum, and minimum ($\mu g/m^3$) of concentration of air pollutants (SO_2 and TSP) in fertilizer factory stack emission from August 2015 to November 2016.

Pollutants	Year Parameter	2015					2016										
		Aug.	Sep.	Oct.	Nov.	Dec.	Jan.	Feb.	Mar.	Apr.	May	Jun.	Jul.	Aug.	Sep.	Oct.	Nov.
SO_2	Median	270	324a	330	324a	320	375a	375a	377a	513	393b	372	392b	419	525	564c	564c
	P25	252	316	309	313	303	352	337	361	405	371	345	378	387	496	504	516
	P75	314	356	344	348	332	388	403	409	563	416	396	412	453	552	605	596
	Min.	219	278	282	8	243	325	317	103	370	299	28	266	227	444	9	492
	Max.	348	367	379	387	379	437	412	432	757	445	417	421	538	576	1149	687
TSP	Median	37*	37*	37*	37*	36*	36*	37*	37*	37*	36*	37*	37*	34	37*	37*	37*
	P25	37	37	32	37	27	31	37	37	37	32	37	37	31	36	36	37
	P75	37	37	37	38	37	38	37	37	37	37	37	37	37	37	37	37
	Min.	37	37	27	29	22	25	36	32	32	28	38	32	23	29	33	36
	Max.	37	37	40	42	38	40	37	38	38	37	37	37	27	37	37	37

In the same raw: the same letter, and * (for SO_2 and PM_{10} respectively) indicates no significantly different among month (*k independent Kruskal–Wallis, P>0.05*) in each year separately. Raw data obtained from EEAA.

31

3.I.b Soil, plant, and water pollutants

Heavy metals concentrations in soil, plant, and water samples collected from polluted sites were significantly different to the that collected from control site (Fig. 8-10). The concentration of Cu in collected soil, and water samples was always below 1 mg/kg dry weight of soil, and mg/ L of water. Copper contents was significantly higher only in soil sample from Site A in comparison to control site.

The highest concentration of Cd was from site A. Usually, there were no significant differences in Cd and Pb concentration from site B, and C comparing to control (*Tukey's-b,* ANOVA, $p > 0.05$).

Phosphate concentration in soil, plant, and water samples was significantly higher comparing to the control site. Mean values at sites A, B, and C were, respectively, 16.5, 6.0, and 4.8 (for soil samples); 1.3, 2.6, and 2.1 (for plant samples); 1.8, 1.3, and 1.3 (for water samples) times higher than control. All mean values differed significantly among each other in soil, and plant samples (*Tukey's-b,* ANOVA, $p > 0.05$). Sulphate concentrations in collected soil samples in all experimental sites were significantly higher than in control samples. The mean values in soil samples decreased gradually with distance from the contamination source. But the mean values of its concentration have no definite trend. Sulphate concentrations at sites A, B, and C were, respectively, 8.8, 5.7, and 8.0 (for soil samples); 1.4, 0.84, and 1.2 (for plant samples); 2.5, 2.1, and 1.4 (for water samples) time higher significantly in relation to the control site (*Tukey's-b,* ANOVA, $p > 0.05$).

Assessment of the overall relationship among sampling sites distance and mean concentration of heavy metals, sulphates, and phosphates in soil, plant, and water samples revealed strong negative correlation except for Cu concentration in plant, and water samples; Zn, and PO_4^{3-} in plant samples. (Table 4).

Results

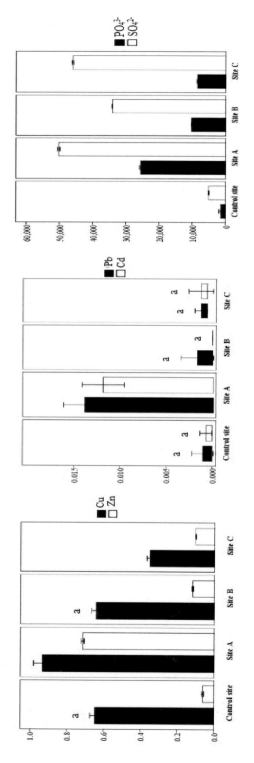

Fig. 8. Concentrations of heavy metals (Cu, Zn, Pb, and Cd), PO$_4$$^{3-}$, and SO$_4$$^{2-}$ (mean ± SE; mg/kg dry weight) in soil samples collected at control and polluted sites (A-C) located at different distances from fertilizer factory.

The same letters denote homogeneous groups (no significant difference occur among groups); *(Tukey's-b, $p > 0.05$)*. Statistical analysis was made for each pollutant separately.

34

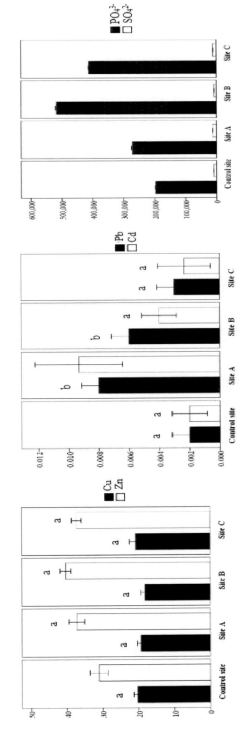

Fig. 9. Concentrations of heavy metals (Cu, Zn, Pb, and Cd), PO$_4$$^{3-}$, and SO$_4$$^{2-}$ (mean ± SE; mg/kg dry weight) in plant samples collected at control and polluted sites (A–C) located at different distances from the fertilizer factory.

The same letters denote homogeneous groups (no significant difference occur among groups); (*Tukey's-b*, $p > 0.05$). Statistical analysis was made for each pollutant separately.

35

Results

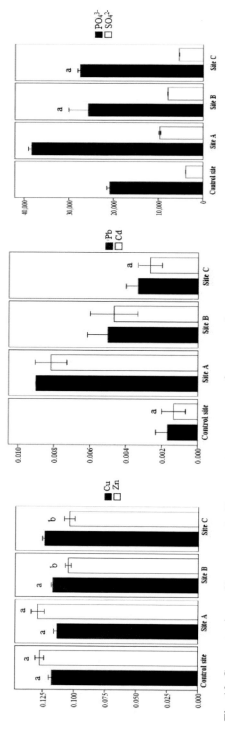

Fig. 10. Concentrations of heavy metals (Cu, Zn, Pb, and Cd), PO_4^{3-}, and SO_4^{2-} (mean ± SE; mg/L) in water samples collected at control and polluted sites (A-C) located at different distances from the fertilizer factory.

The same letters denote homogeneous groups (no significant difference occur among groups); (*Tukey's-b*, $p > 0.05$). Statistical analysis was made for each pollutant separately.

36

Table 4. Pearson's correlation coefficient among concentrations of heavy metals (Cu, Zn, Pb, and Cd), PO_4^{3-}, and SO_4^{2-} of soil, plant, and water samples collected at control and polluted sites (A-C) located at different distances from fertilizer factory.

Pollutants	SS[#]	Regression analysis	Type of equation	R^2	r
Cu	Soil	$y= 0.006x^2 - 0.157x + 1.072$	Polynomial	1	-0.82**
	Plant	$y= 17.971 \ x^{0.104}$	Power	0.76	0.41
	Water	$y= 0.114 \ e^{0.010 \ x}$	Exponential	0.99	0.93**
Zn	Soil	$y= 0.059x^2 - 0.537x + 1.188$			-0.88**
	Plant	$y= -0.853x^2 + 6.613x + 29.640$	Polynomial	1	0.007
	Water	$y= 0.002x^2 - 0.018x + 0.146$			-0.85**
Pb	Soil	$y= 0.001x^2 - 0.009x + 0.022$			-0.88**
	Plant	$y=-0.003 \ln (x) + 0.0093$	Logarithmic	0.94	-0.99**
	Water	$y= 0.0005x^2 - 0.004x + 0.013$			-0.94**
Cd	Soil	$y= 0.001x^2 - 0.013x + 0.025$	Polynomial	1	-0.87**
	Plant	$y= 0.0005x^2 - 0.005x + 0.013$			-0.85**
	Water	$y= 0.011 \ e^{0.284 \ x}$	Exponential	9.72	-0.94**
PO_4^{3-}	Soil	$y= 1416.7x^2 - 13317x + 37600$			-0.91**
	Plant	$y= -31711x^2 + 250845x + 49867$			0.57
	Water	$y= 1011.1x^2 - 9210.9x + 46866$	Polynomial	1	-0.75**
SO_4^{2-}	Soil	$y= 2461.7x^2 - 18072x + 66140$			-0.24
	Plant	$y= 720.811x^2 - 5472.20x + 16995$			-0.33
	Water	$y= 29.667x^2 - 1096.7x + 10889$			-0.99**

* significant at $p < 0.05$; ** significant at $p < 0.001$. #SS Samples Source.

A cluster analysis using Ward's method revealed slightly dissimilar patterns for soil, plant, and water samples, however, the general tendency was similar (Fig 11 and 12). The concentrations of pollutants (heavy metals (Cu, Zn, Pb, and Cd), PO_4^{3-}, and SO_4^{2-}) were highly similar in site B, and C in soil, plant, and water samples. Pollutants concentration were linked among site A, B, and C (in soil samples); sites B and C (in plant samples); control site, sites B, and C in water samples. Concentration of pollutants created a separate cluster in control site (soil samples), and site A (water samples). The cluster of control site and polluted sites had relatively high distances (Fig. 11).

A hierarchical cluster analysis of concentrations of pollutants (heavy metals (Cu, Zn, Pb, and Cd), PO_4^{3-}, and SO_4^{2-}) (Fig. 12) enabled to create separate clusters in plant samples of control site, and polluted sites (A-C). Pollutants concentration in soil, and water samples of control site, and polluted sites (A-C) was almost the same (Fig. 12).

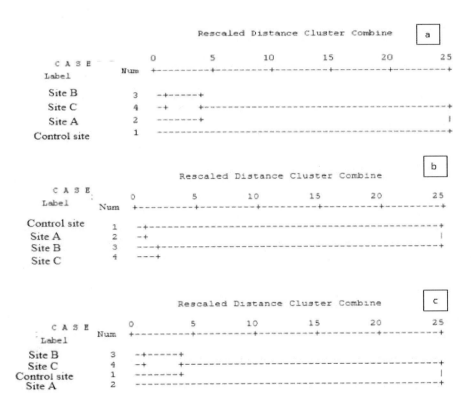

Fig. 11. Dendrogram of the cluster analysis (using Ward's Method) applied for concentration of soil (a), plant (b), and water (c) pollutants (heavy metals (Cu, Zn, Pb, and Cd), PO_4^{3-}, and SO_4^{2-}) collected at control site and polluted sites (A-C) located at different distances from the fertilizer factory.

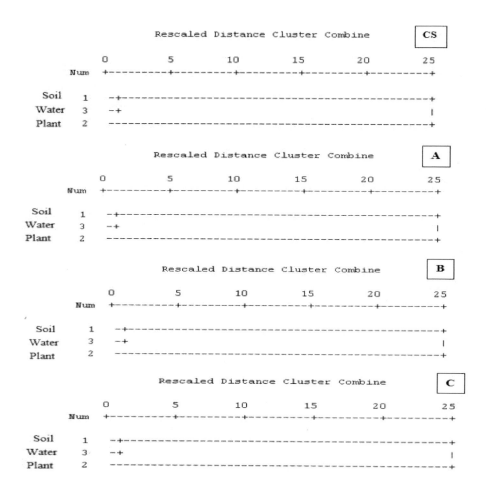

Fig. 12. Dendrogram of the cluster analysis (using Ward's Method) applied to show accumulation difference of pollutants (heavy metals (Cu, Zn, Pb, and Cd), PO_4^{3-}, and SO_4^{2-}) in soil, plant, and water samples collected from control site(CS), and polluted sites (A, B, C) located at different distances from the fertilizer factory.

Chi-square (x^2) test using cross tabulation was performed to examine independency among stack emission pollutants of AZFC (SO_2, and TSP) in air, and pollutants in soil, plant, and water (heavy metals (Cu, Zn, Pb, and Cd), PO_4^{3-}, and SO_4^{2-}) (Table 5). There was a strong positive correlation among air pollutants and SO_4^{2-} in soil, plant, and water (0.7, 0.3, and 0.7, respectively). The synergetic effect of air pollutants concentration on soil, plant, and water pollutants concentration was found ($p < 0.05$) (Table 5). There was no relation among the air pollutants concentration and copper in plant samples. Also, for water samples, there was no relation among the air pollutants concentration and lead concentration ($p > 0.05$).

Chi-square (x^2) test revealed the relationship among heavy metals concentration in soil, plant, and water and PO_4^{3-}, and SO_4^{2-} in the same samples collected from polluted area around AZFC. The significant level of x^2 was lower than 0.05, the correlations were almost strong positive among pollutants concentration (Table 6).

The results showed that there was a positive correlation (weak to strong) among soil pollutants (heavy metals (Cu, Zn, Pb, and Cd), PO_4^{3-}, and SO_4^{2-}) concentration, and concentration of these pollutants in plant, and water samples ($p < 0.05$) (Table 7).

Table 5. Chi-square (x^2) using cross tabulation was performed to examine independency among air pollutants concentration (SO_2 and TSP) in Abu-Zaabal stack emission and soil, plant, and water pollutants concentration (heavy metals (Cu, Zn, Pb, and Cd), PO_4^{3-} and SO_4^{2-}) in samples collected from polluted sites around the fertilizer factory.

	Pollutants	X^2	LR	*df*	LLA	*df*	R	r	T	*p* value
	Cu	6.0	6.5	4	1.8	1	0.9	1.0	4.8	< 0.0001
	Zn	6.0	6.5	4	0.5	1	0.5	0.5	4.8	< 0.0001
Soil	Pb	6.0	6.5	4	2.0	1	1.0	1.0	4.8	< 0.0001
	Cd	6.0	6.5	4	0.5	1	0.5	0.5	4.8	< 0.0001
	PO_4^{3-}	6.0	6.5	4	1.7	1	0.9	1.0	4.8	< 0.0001
	SO_4^{2-}	6.0	6.5	4	1.1	1	0.7	0.5	4.8	< 0.0001
	Cu	6.0	6.5	4	0.3	1	0.3	0.5	4.8	< 0.0001
	Zn	6.0	6.5	4	0.5	1	0.5	0.5	4.8	< 0.0001
plant	Pb	6.0	6.5	4	0.5	1	0.5	0.5	4.8	< 0.05
	Cd	6.0	6.5	4	1.9	1	0.9	1.0	4.8	< 0.0001
	PO_4^{3-}	6.0	6.5	4	1.2	1	0.9	1.0	4.8	< 0.0001
	SO_4^{2-}	6.0	6.5	4	0.2	1	0.3	0.5	4.8	< 0.05
	Cu	3.0	3.8	2	0.0	1	0.0	0.0	1.2	>0.05
	Zn	6.0	6.5	4	0.8	1	0.6	0.5	2.4	< 0.0001
Water	Pb	3.0	3.8	2	0.0	1	0.0	0.0	1.2	>0.05
	Cd	6.0	6.5	4	1.9	1	0.9	1.0	4.8	< 0.0001
	PO_4^{3-}	6.0	6.5	4	1.4	1	0.8	0.5	4.8	< 0.0001
	SO_4^{2-}	6.0	6.5	4	1.7	1	0.7	0.5	4.8	< 0.0001

X^2, LR, LLA, R, r, and T represent of Person's chi-square, Likelihood Ratio, Linear by Linear Association, Person's R, correlation, and T of Lambda respectively. x^2 revealed was made for each pollutant (Cu, Zn, Pb, and Cd), PO_4^{3-}, and SO_4^{2-}), and environment component (soil, water, and plant) separately.

Table 6. Chi-square (x^2) using cross tabulation was performed to examine independency among heavy metals concentration (Cu, Zn, Pb, and Cd) and PO_4^{3-}, and SO_4^{2-} in soil, plant, and water samples collected from polluted sites around the fertilizer factory.

	Pollutants	X^2	LR	*df*	LLA	*df*	R	r	T	*p* value
Soil	PO_4^{3-}	72.7	43.2	66	9.2	1	0.9	0.6	5.2	< 0.0001
	SO_4^{2-}	72.4	43.1	66	2.9	1	0.5	0.6	6.0	< 0.0001
Plant	PO_4^{3-}	96.1	51.3	88	0.4	1	0.5	0.3	6.8	< 0.0001
	SO_4^{2-}	96.2	51.4	88	0.4	1	0.5	0.3	6.8	< 0.0001
Water	PO_4^{3-}	72.1	44.2	66	8.9	1	0.9	0.8	6.0	< 0.0001
	SO_4^{2-}	72.3	44.4	66	9.9	1	0.9	0.9	6.0	< 0.0001

X^2, LR, LLA, R, r, and T represent of Person's chi-square, Likelihood Ratio, Linear by Linear Association, Person's R, correlation, and T of Lambda respectively. x^2 revealed was made for each pollutant (PO_4^{3-}, and SO_4^{2-}), and environment component (soil, water, and plant) separately.

Table 7. Chi-square (x^2) using cross tabulation was performed to examine independency among soil pollutants concentration (heavy metals concentration (Cu, Zn, Pb, and Cd) and PO_4^{3-}, and SO_4^{2-}) and these pollutants in plant, and water samples collected from polluted sites around the fertilizer factory.

	Pollutants	X^2	LR	*df*	LLA	*df*	R	r	T	*p* value
	Cu	108.0	54.0	99	1.4	1	-0.3	-0.31	8.5	< 0.0001
	Zn	120.0	56.0	110	0.4	1	0.1	0.6	9.2	< 0.0001
plant	Pb	59.0	38.0	48	6.1	1	0.7	0.6	5.8	< 0.0001
	Cd	47.0	29.1	35	8.3	1	0.8	0.5	2.8	< 0.05
	PO_4^{3-}	132.2	59.2	121	0.001	1	0.008	0.3	17.0	< 0.0001
	SO_4^{2-}	132.1	59.3	121	4.4	1	0.6	0.7	17.0	< 0.0001
	Cu	84.1	47.5	77	9.1	1	-0.9	-0.76	6.2	< 0.0001
	Zn	108.1	54.1	99	3.2	1	0.5	0.1	8.5	< 0.0001
Water	Pb	43.5	34.1	36	8.8	1	0.8	0.7	4.8	< 0.0001
	Cd	47.2	29.1	35	7.4	1	0.8	0.4	2.8	< 0.05
	PO_4^{3-}	132.5	59.1	121	10.1	1	0.9	0.8	11.5	< 0.0001
	SO_4^{2-}	132.2	59.1	121	5.7	1	0.7	0.8	3.3	< 0.05

X^2, LR, LLA, R, r, and T represent of Person's chi-square, Likelihood Ratio, Linear by Linear Association, Person's R, correlation, and T of Lambda respectively. x^2 revealed was made for each pollutant ((Cu, Zn, Pb, and Cd), PO_4^{3-}, and SO_4^{2-}), and environment component (water, and plant) separately.

2 Oxidative stress

2.I Reactive oxygen species assay

2.I.a Hydrogen peroxide concentration (H_2O_2)

In the control groups, the H_2O_2 concentration was significantly lower in homogenates of brain than in thoracic muscle and gut of females of *A. thalassinus*. In males, females, and nymph from polluted sites median values of the H_2O_2 concentration were significantly higher in insects from the polluted sites than in those from control site ($p < 0.05$) (Fig 13). The highest values of H_2O_2 concentration was found in the gut of females collected from site A. Significant differences among sexes (male, and female), and nymph regarding H_2O_2 concentration in thoracic muscles were observed in the individuals from all the polluted site.

2.I.b Superoxide anion radical ($O_2^{\cdot-}$)

The $O_2^{\cdot-}$ production rate was significantly lower in homogenates of brain than in thoracic muscle and gut of males of *A. thalassinus* collected from control site. In males, females, and nymph median values of the $O_2^{\cdot-}$ production rate was significantly higher in insects from the polluted sites than in those from control site ($p < 0.05$) (Fig 14). In the brain of females from site A the highest values of $O_2^{\cdot-}$ production rate was observed. Significant differences among sexes (male, and female), and nymph regarding $O_2^{\cdot-}$ production rate in brain, thoracic muscles, and gut were observed in individuals from all polluted site.

A correlation analysis among H_2O_2 concentration, and the distance from fertilizer factory, revealed the most significant relationships in gut (Table 8). In female, and nymph gut a correlation at $p<0.001$ among the distance from the source of contamination and H_2O_2 concentration were shown. In case of male gut, a significant correlation at $p<0.05$ was observed. Also, the correlation among $O_2^{\cdot-}$ production rate and distance away from AZFC showed that 0.001 significant level occur in thoracic muscles and gut, in male, female, and nymph. However, the interaction analysis showed significant influence of distance, sex and tissue on all comet parameters (Table 13).

A cluster analysis using Ward's method showed slightly similar general patterns for males, females, and nymph (Fig 15, 16 and 17). The ROS concentration (H_2O_2, and $O_2^{\cdot-}$) was highly similar in thoracic muscles of males collected at sites B, and site C. In female and nymph, ROS was similar in thoracic muscles at control site and site C. The ROS in brain tissue of males from control site was also linked with those from sites A and B. In thoracic muscles tissues from control site was linked with sites B and C. In female, control site linked with sites C and B in thoracic muscles tissue. In gut homogenates of nymph, control site linked with sites A and B. Separate cluster was observed in males from site C, and A in brain, and thoracic muscle, respectively (Fig. 15). A hierarchical cluster analysis of ROS in female thoracic muscles tissue also enabled to create separate clusters (Fig. 16). In brain, and gut tissues in nymph also enabled to create separate clusters at sites B, and C, respectively (Fig. 17).

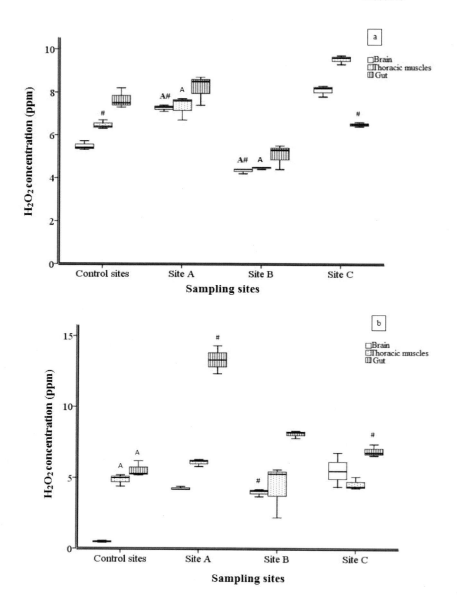

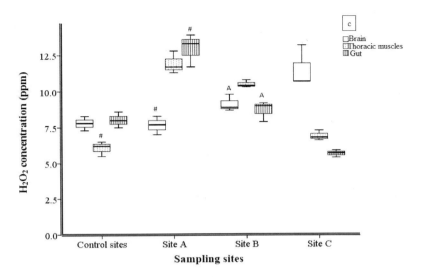

Fig. 13. Hydrogen peroxide (H₂O₂) concentration, expressed as median, percentile deviation (P25 and P75 -boxes), and min-max values of H_2O_2 concentration obtained from brain, thoracic muscles, and gut homogenates of males (a), females (b), and nymph (c) of *Aiolopus thalassinus* collected at different distances from the fertilizer factory.

Median values marked with different small letters are significantly different among control and polluted sites (A-C) located at different distances from the fertilizer factory. Median values marked with different capital letters are significantly different among tissues (Kruskal-Wallis revealed, $p < 0.05$). # denote no significant differences among males, females, and nymph in each case separately (Kruskal-Wallis revealed, $p < 0.05$).

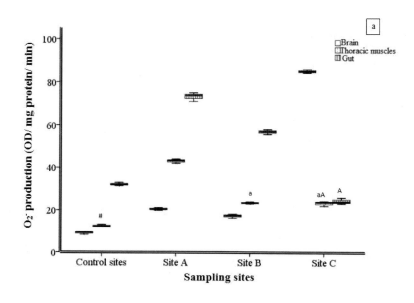

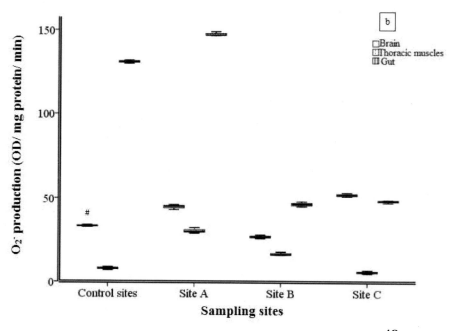

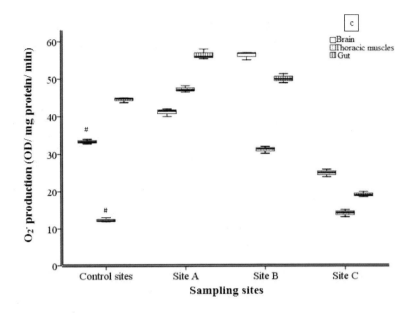

Fig. 14. Superoxide anion (O₂⁻) production rate, expressed as median, percentile deviation (P25 and P75 -boxes), and min-max values of H_2O_2 concentration obtained from brain, thoracic muscles, and gut homogenates of males (a), females (b), and nymph (c) of *Aiolopus thalassinus* collected at different distances from the fertilizer factory.

Median values marked with different small letters are significantly different among control and polluted sites (A-C) located at different distances from the fertilizer factory. Median values marked with different capital letters are significantly different among tissues (Kruskal-Wallis revealed, $p < 0.05$). # denote no significant differences among males, females, and nymph in each case separately (Kruskal-Wallis revealed, $p < 0.05$).

Table 8. Pearson's correlation coefficient among reactive oxygen species (ROS) (hydrogen peroxide concentration (H_2O_2), and superoxide anion production level (O_2^-)) in brain, thoracic muscles, and gut cells of males, females, and 5th instar of *Aiolopus thalassinus* and the distance from the fertilizer factory.

ROS	Tissue Sex	Brain Regression analysis	r	Thoracic muscles Regression analysis	r	Gut Regression analysis	r
H_2O_2	Male	$y= 3.35x^2 - 12.95x + 16.90$	+0.21	$y= 4.1x^2 - 15.4x + 18.91$	+0.42	$y= 2.21x^2 - 9.81x + 16.10$	-0.51
	Female	$y= 0.74x^2 - 2.31x + 5.77$	+0.59	$y= -1.60 \ln (x) + 6.25$	-0.51	$y= 1.85x^2 - 10.65x + 22.10$	-0.92**
	Nymph	$y= 0.31x^2 + 0.31x + 7.11$	+0.88**	$y= -1.15x^2 + 2.15x + 10.71$	-0.95**	$y= 0.51x^2 - 5.81x + 18.61$	-0.97**
O_2^-	Male	$y= 35.25x^2 - 108.75x + 94.11$	+0.84**	$y= 10.05x^2 - 49.95x + 83.21$	-0.86**	$y= -8.15x^2 + 7.75x + 74.41$	-0.98**
	Female	$y= 21.51x^2 - 82.20x + 106.11$	+0.28	$y= 106.81x^2 - 18.20x + 97.41$	-0.99**	$y= 51.81x^2 - 255.20x + 351.41$	-0.85**
	Nymph	$y= -23.81x^2 + 84.20x - 20.21$	-0.51	$y= 0.51x^2 - 14.51x + 62.41$	-0.99**	$y= -12.51x^2 + 31.52x + 37.21$	-0.93**

* significant at p<0.05; ** significant at p<0.001

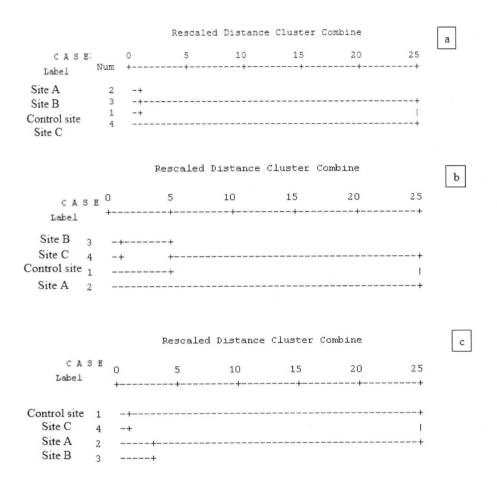

Fig. 15. Dendrogram of the cluster analysis (using Ward's Method) applied for reactive oxygen species (ROS) (hydrogen peroxide concentration, and superoxide anion production rate) in brain (a), thoracic muscles (b), and gut (c) homogenates of males of *Aiolopus thalassinus*, which were collected at control site and polluted sites (A-C) located at different distances from the fertilizer factory.

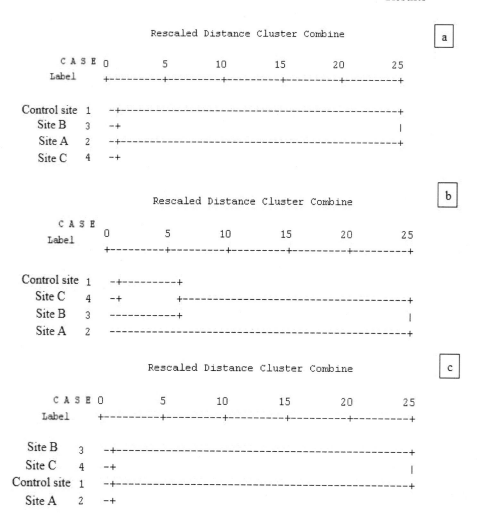

Fig. 16. Dendrogram of the cluster analysis (using Ward's Method) applied for reactive oxygen species (ROS) (hydrogen peroxide concentration, and superoxide anion production rate) in brain (a), thoracic muscles (b), and gut (c) homogenates of females of *Aiolopus thalassinus*, which were collected at control site and polluted sites (A-C) located at different distances from the fertilizer factory.

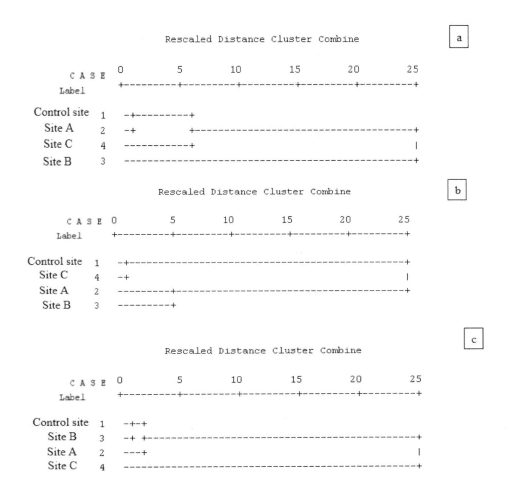

Fig. 17. Dendrogram of the cluster analysis (using Ward's Method) applied for reactive oxygen species (ROS) (hydrogen peroxide concentration, and superoxide anion production rate) in brain (a), thoracic muscles (b), and gut (c) homogenates of 5th instar of *Aiolopus thalassinus*, which were collected at control site and polluted sites (A-C) located at different distances from fertilizer factory.

2.II Oxidative damage assays

A negative control of oxidative damage was performed on insect *A. thalassinus* reared under laboratories conditions for three generations and compared with the control specimens collected from control sites. There was no significant difference among level of DNA damage in the tissues of untreated individuals (negative control), and insects from control site ($p > 0.05$).

Wald-wolfowitz run revealed was performed to revealed randomness of data obtained from experiments, the result proved randomness of data ($p > 0.05$).

2.II.a DNA damage assay-Comet assay

The results showed that, the level of the DNA damage was higher in insects from polluted sites comparing to that from control site (Fig. 18-22). The highest values of tail length (TL) and % of DNA (TDNA) were found in the gut of males from site C (Fig. 18 and 20). Significant differences among both sexes, and nymph regarding TDNA parameter in brain were observed in individuals from all the polluted site (Fig. 20).

In the control groups the DNA damage, expressed as TL, was significantly lower in cells of thoracic muscles than in brain and gut of males, and nymph of *A. thalassinus*, while in females no significant differences were observed in the examined tissues. In male, female and nymph, median values of TL parameter were significantly higher in insects from the polluted sites than in those from control site ($p < 0.05$), and did not have a close relationship with the distance from the fertilizer factory (Fig. 18).

The values of Olive Moment (OM) parameter were also the lowest in tissues of insects collected from control site. However, the statistical analysis revealed significant differences among males, females, and nymphs from all sites, and in almost all tissues. Median OM values were homogenous in brain of males, females and nymphs collected at from site B, and gut tissue of both sexes from site C (Fig. 21). The highest values of Olive Moment were observed in the gut of males from sites C and B, while in females, it was detected in the gut of individuals collected from site A (Fig. 21). The highest OM value in nymphs was also detected in the gut of individuals collected from site C.

The highest percentage of cells with DNA damage (as % severed cells) was observed, in both males, females and nymphs collected from site A, in gradient waiver from males, and nymphs then females. (Fig. 22). The results showed that tail moment had the highest values in the gut of males, and nymphs from sites A and C, while in females, it was in the gut of individuals inhabiting only site A. (Fig. 19).

Correlation analysis among comet parameters and the distance from AZFC was done (Table 9). In case of male, female, and nymph thoracic muscles, a significant correlation at $p<0.001$ was observed only for % severed cells parameter. Moreover, significant relationships among the distance from fertilizer factory and TDNA or OM in thoracic muscles of males, and nymphs were found at $p<0.05$. Highly significant negative correlation (at $p<0.001$) was revealed among the distance and % severed cells in all tissues of both sexes; and for gut tissue in nymph. In female thoracic muscles, a correlation at $p<0.001$ were shown among the distance from the source of contamination and

55

TL or TDNA or OM or % severed cells parameters. This parameter seems to be the most useful for biomonitoring of genotoxicity of industrial fertilizer pollutants (Table 9). However, the interaction analysis showed significant influence of distance, sex and tissue on all comet parameters (Table 10). Also, interaction among distance, developmental stage, and tissues had a significant effect on all comet parameters (Table 11).

A cluster analysis using Ward's method revealed slightly dissimilar patterns for males, females and nymphs, however, the general tendency was similar (Fig. 23-25). The level of DNA damage was highly similar in brain as well as thoracic muscle of males and nymphs collected at sites B and C (Fig. 23). The level of DNA damage in this tissue of males, and nymphs from site A was also linked with those from sites B and C. Males, and nymphs from control site created a separate cluster. Control site cluster and polluted sites cluster had relatively high distances. A hierarchical cluster analysis of DNA damage in female tissues also enabled to create separate clusters (Fig. 24). The level of DNA damage in the brain as well as in the gut of females from sites B and C was almost the same. The DNA damage in the female thoracic muscles was almost the same in the individuals from sites A and C. Females from control site created a separate cluster unlike the cluster of polluted sites.

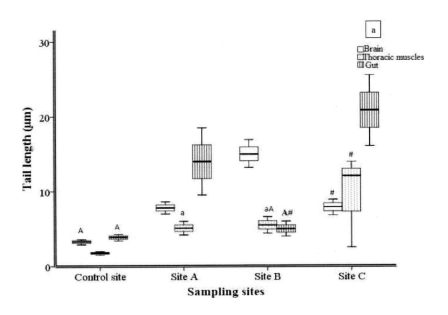

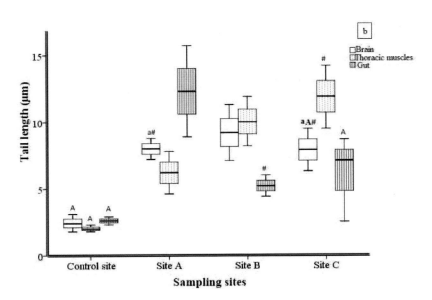

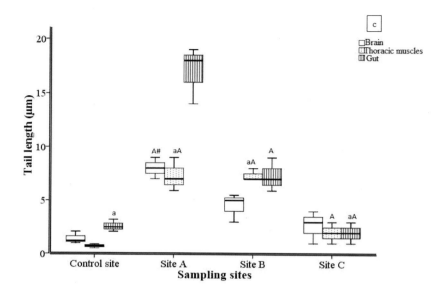

Fig. 18. Tail length (TL; μm), expressed as median and percentile deviation (P25 and P75), of comets obtained from brain, thoracic muscles, and gut cells of males (a), females (b), and 5th instar (c) of *Aiolopus thalassinus*, which were collected at different distances from the fertilizer factory.

Median values marked with different small letters are significantly different among control and polluted sites located at different distances from the fertilizer factory. Median values marked with different capital letters are significantly different among tissues (Kruskal-Wallis revealed, $p < 0.05$). # denote no significant differences among males, females, and nymph in each case separately (Kruskal-Wallis revealed, $p < 0.05$).

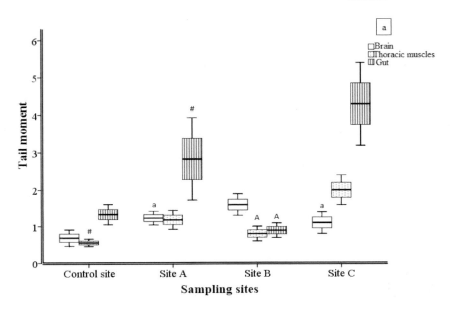

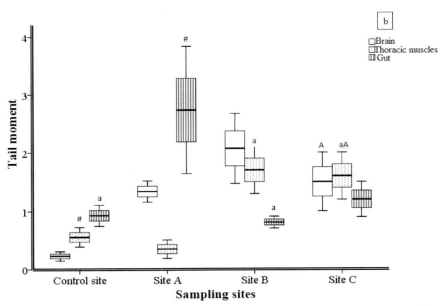

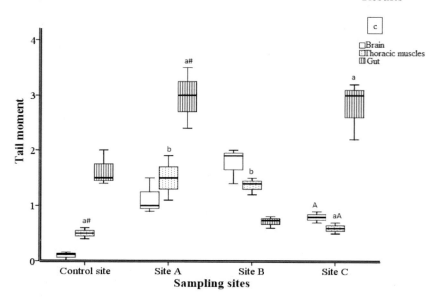

Fig. 19. Tail moment (TM), expressed as median and percentile deviation (P25 and P75), of comets obtained from brain, thoracic muscles, and gut cells of males (a), females (b), and 5th instar (c) of *Aiolopus thalassinus*, which were collected at different distances from fertilizer factory.

Median values marked with different small letters are significantly different among control and polluted sites located at different distances from the fertilizer factory. Median values marked with different capital letters are significantly different among tissues (Kruskal-Wallis revealed, $p < 0.05$). # denote no significant differences among males, females, and nymph in each case separately (Kruskal-Wallis revealed, $p < 0.05$).

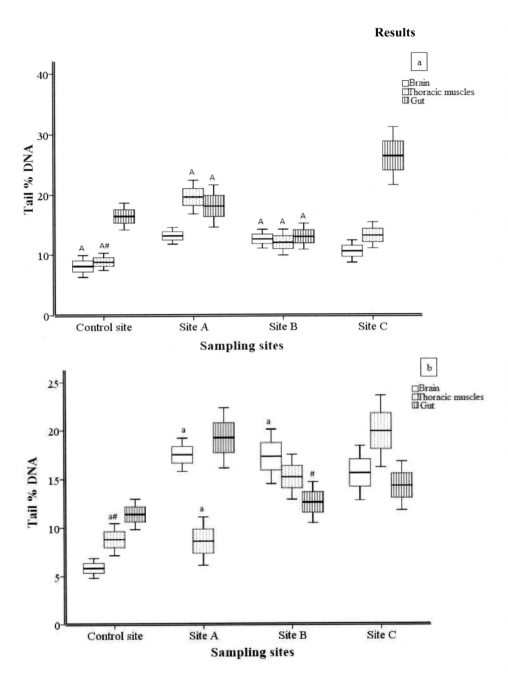

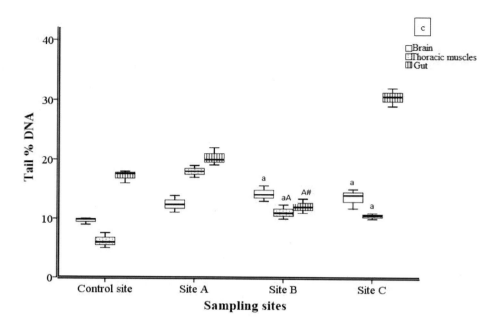

Fig.20. Tail % DNA, expressed as median and percentile deviation (P25 and P75), of comets obtained from brain, thoracic muscles, and gut cells of males (a), females (b), and 5th instar (c) of *Aiolopus thalassinus*, which were collected at different distances from fertilizer factory.

Median values marked with different small letters are significantly different among control and polluted sites located at different distances from the fertilizer factory. Median values marked with different capital letters are significantly different among tissues (Kruskal-Wallis revealed, $p < 0.05$). # denote no significant differences among males, females, and nymph in each case separately (Kruskal-Wallis revealed, $p < 0.05$).

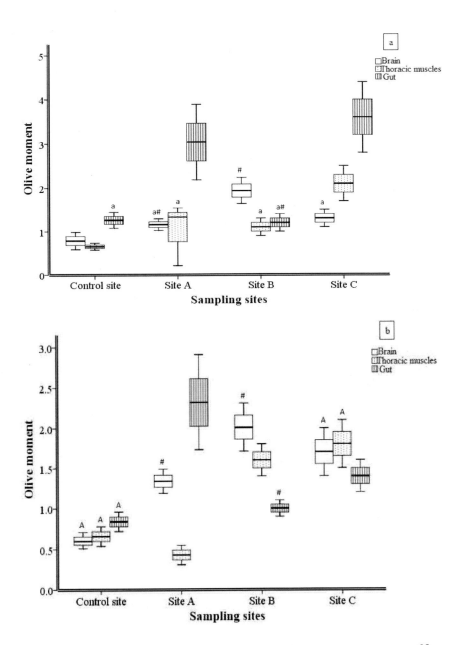

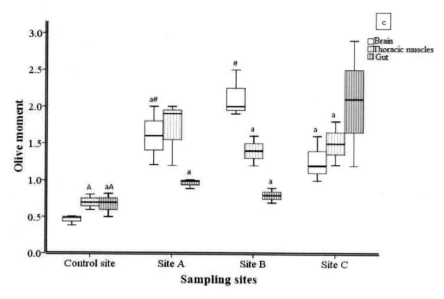

Fig. 21. Olive moment (OM), expressed as median and percentile deviation (P25 and P75), of comets obtained from brain, thoracic muscles, and gut cells of males (a), females (b), and 5th nymphal instar (c) of *Aiolopus thalassinus*, which were collected at different distances from fertilizer factory.

Median values marked with different small letters are significantly different among control and polluted sites located at different distances from the fertilizer factory. Median values marked with different capital letters are significantly different among tissues (Kruskal-Wallis revealed, $p < 0.05$). # denote no significant differences among males, females, and nymph in each case separately (Kruskal-Wallis revealed, $p < 0.05$).

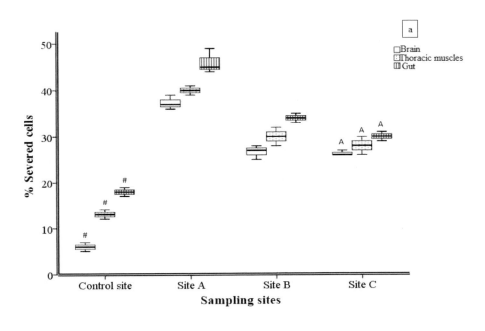

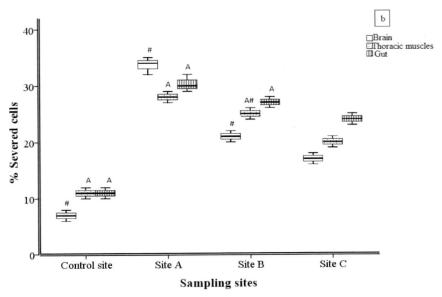

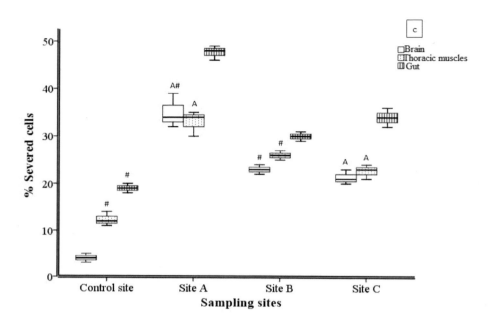

Fig. 22. The % severed cells, expressed as median and percentile deviation (P25 and P75), of comets obtained from brain, thoracic muscles, and gut cells of males (a), females (b), and 5th instar (c) of *Aiolopus thalassinus*, which were collected at different distances from fertilizer factory.

Median values marked with different small letters are significantly different among control and polluted sites located at different distances from the fertilizer factory. Median values marked with different capital letters are significantly different among tissues (Kruskal-Wallis revealed, $p < 0.05$). # denote no significant differences among males, females, and nymph in each case separately (Kruskal-Wallis revealed, $p < 0.05$).

Table 9. Pearson's correlation coefficient among comet parameters (tail length, %DNA in tail, olive moment, tail moment, and % severed cells) in brain, thoracic muscles, and gut homogenates of males, females, and 5th instar of *Aiolopus thalassinus* and the distance from fertilizer factory.

Comet parameters	Tissue Sex	Brain Regression analysis	r	Thoracic muscles Regression analysis	r	Gut Regression analysis	r
Tail length	Male	$y=-0.71x^2 + 2.86x - 1.37$	+0.008	$y=0.31x^2 - 0.89x + 1.09$	+0.50	$y=1.24x^2 - 4.63x + 4.79$	+0.38
	Female	$y=-0.12x^2 + 0.49x + 0.43$	-0.02	$y=0.09x^2 + 0.66x + 0.05$	+0.81**	$y=0.45x^2 - 2.06x + 2.84$	-0.65
	Nymph	$y=0.01x^2 - 0.15x + 0.84$	-0.98	$y=-0.03x^2 + 0.20x + 0.42$	-0.70	$y=0.05x^2 - 0.62x + 1.97$	-0.98
Tail moment	Male	$y=-0.19x^2 + 0.75x + 0.86$	-0.17	$y=0.72x^2 - 2.61x + 3.32$	+0.60	$y=3.78x^2 - 14.38x + 14.53$	+0.37
	Female	$y=-x^2 + 4x - 1$	+0.13	$y=1.39\ln(x) + 0.57$	+0.75*	$y=1.46x^2 - 7.32x + 9.71$	-0.63
	Nymph	$y=-0.11x^2 + 0.72x + 0.28$	-0.13	$y=-0.90x^2 + 0.56x + 0.53$	-0.51	$y=0.32x^2 - 2.08x + 4.16$	+0.32
%DNA in tail	Male	$y=-0.76x^2 + 1.72x + 12.31$	-0.61	$y=4.40x^2 - 20.80x + 36.10$	-0.67*	$y=5.24x^2 - 16.82x - 29.70$	+0.54
	Female	$y=-0.75x^2 + 2.05x + 16.20$	-0.35	$y=-0.43x^2 + 6.87x + 3.19$	+0.88**	$y=4.16x^2 - 19.11x + 34.19$	-0.57
	Nymph	$y=-0.42x^2 + 2.54x + 11.78$	-0.55	$y=0.34x^2 - 3.66x + 20.33$	-0.96	$y=11.45x^2 - 40.05x + 47.7$	+0.65

* significant at $p < 0.05$; ** significant at $p < 0.001$.

Continue table 6. Pearson's correlation coefficient among comet parameters (tail length, %DNA in tail, olive moment, tail moment, and % severed cells) in brain, thoracic muscles, and gut homogenates of males, females, and 5th instar of *Aiolopus thalassinus* and the distance from fertilizer factory.

Comet parameters	Tissue Sex	Brain		Thoracic muscles		Gut	
		Regression analysis	r	Regression analysis	r	Regression analysis	r
Olive moment	Male	$y= -0.71x^2 + 2.91x - 1.04$	+0.14	$y= 0.61x^2 - 2.07x + 2.79$	+0.69*	$y= 2.12x^2 - 8.20x + 9.12$	+0.19
	Female	$y= -0.49x^2 + 2.14x - 0.31$	+0.42	$y= -0.48x^2 + 2.62x - 1.71$	+0.88**	$y= 0.86x^2 - 3.90x + 5.36$	-0.59
	Nymph	$y= -0.08x^2 + 0.70x + 0.59$	+0.56	$y= -0.03x^2 + 0.23x + 1.00$	+0.04	$y= 0.09x^2 - 0.41x + 1.34$	+0.82
% Severed cells	Male	$y= 3x^2 - 17x + 50$	-0.86**	$y= 4x^2 - 22x + 58$	-0.90**	$y= -12.92\ln(x) + 43.71$	-0.94**
	Female	$y= 3x^2 - 19x + 48$	-0.94**	$y= -2x^2 + 4x + 26$	-0.96**	$y= -x^2 + x + 30$	-0.93**
	Nymph	$y= 0.81x^2 - 8.20x + 41.41$	-0.95	$y= 0.10x^2 - 1.90x + 31.8$	-1.00**	$y= 2.03x^2 - 16.61x + 60.61$	-0.68

* significant at $p < 0.05$; ** significant at $p < 0.001$.

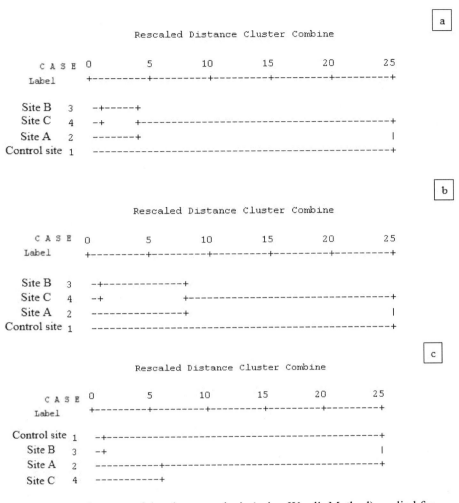

Fig. 23. Dendrogram of the cluster analysis (using Ward's Method) applied for biomarkers (DNA damage analysis) in brain (a), thoracic muscles (b), and gut (c) cells of males of *Aiolopus thalassinus*, which were collected at control site and polluted sites (A-C) located at different distances from fertilizer factory.

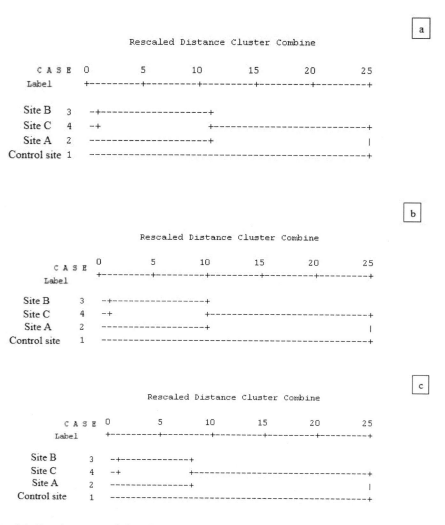

Fig. 24. Dendrogram of the cluster analysis (using Ward's Method) applied for biomarkers (DNA damage analysis) in brain (a), thoracic muscles (b), and gut (c) cells of females of *Aiolopus thalassinus*, which were collected at control site and polluted sites (A-C) located at different distances from fertilizer factory.

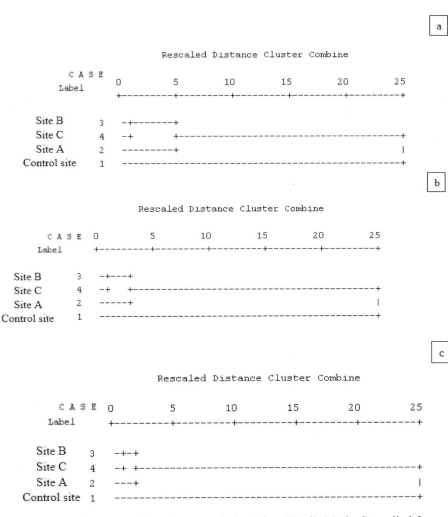

```
                                                              a

                    Rescaled Distance Cluster Combine

          C A S E          0        5       10       15       20       25
          Label            +---------+---------+---------+---------+---------+

     Site B          3    -+-------+
     Site C          4    -+        +-----------------------------------------+
     Site A          2    ---------+                                          |
     Control site    1    ------------------------------------------------------+

                                                              b

                    Rescaled Distance Cluster Combine

         C A S E    0        5       10       15       20       25
         Label      +---------+---------+---------+---------+---------+

    Site B          3    -+---+
    Site C          4    -+    +-----------------------------------------+
    Site A          2    -----+                                          |
    Control site    1    ------------------------------------------------------+

                                                              c

                    Rescaled Distance Cluster Combine

         C A S E    0        5       10       15       20       25
         Label      +---------+---------+---------+---------+---------+

     Site B          3    -+-+
     Site C          4    -+ +-----------------------------------------+
     Site A          2    ---+                                         |
     Control site    1    ------------------------------------------------------+
```

Fig. 25. Dendrogram of the cluster analysis (using Ward′s Method) applied for biomarkers (DNA damage analysis) in brain (a), thoracic muscles (b), and gut (c) cells of 5th nymphal instar of *Aiolopus thalassinus*, which were collected at control site and polluted sites (A-C) located at different distances from fertilizer factory.

Table 10. Generalized Estimating Equation to analyze the interactions among the distance from fertilizer factory, types of tissues, sex on comet parameters (Tail length, Tail DNA, and Olive Tail Moment) in brain, thoracic muscles, and gut cells of males, and females of *Aiolopus thalassinus* collected at different sites located at various distances from fertilizer factory.

Source	Chi-square (χ^2)	df	p value
Distance – sex interaction			
Tail length	6.43	2	<0.05
Tail DNA	10.43	2	<0.05
Olive moment	19.81	2	<0.0001
Tail moment	32.16	2	<0.0001
% severed cells	35.50	2	<0.0001
Distance – tissue interaction			
Tail length	106.29	4	<0.0001
Tail DNA	27.18	4	<0.0001
Olive moment	133.05	4	<0.0001
Tail moment	94.90	4	<0.0001
% severed cells	37.41	4	<0.0001
Sex – tissue interaction			
Tail length	23.67	2	<0.0001
Tail DNA	48.88	2	<0.0001
Olive moment	30.22	2	<0.0001
Tail moment	18.21	2	<0.0001
% severed cells	22.17	2	<0.0001
Distance – tissue – sex interaction			
Tail length	60.12	4	<0.0001
Tail DNA	72.30	4	<0.0001
Olive moment	28.36	4	<0.0001
Tail moment	25.76	4	<0.0001
% severed cells	83.28	4	<0.0001

Table 11. Generalized Estimating Equation to analyze the interactions among the distance from fertilizer factory, types of tissues, developmental stage (adult and nymph) on comet parameters (Tail length, Tail DNA, and Olive Tail Moment) in brain, thoracic muscles, and gut cells of males, females, and 5th instar of *Aiolopus thalassinus* collected at different sites located at various distances from fertilizer factory.

Source	Chi-square (χ^2)	df	p value
Distance – Developmental stage interaction			
Tail length	218.05	2	<0.0001
Tail DNA	141.20	2	<0.0001
Olive moment	3.43	2	>0.05
Tail moment	5.96	2	>0.05
% severed cells	9.78	2	<0.05
Distance – tissue interaction			
Tail length	53.26	4	<0.0001
Tail DNA	466.86	4	<0.0001
Olive moment	68.05	4	<0.0001
Tail moment	22.77	4	<0.0001
% severed cells	11.43	4	<0.05
Developmental stage – tissue interaction			
Tail length	33.37	2	<0.0001
Tail DNA	279.10	2	<0.0001
Olive moment	13.92	2	<0.05
Tail moment	2.23	2	>0.05
% severed cells	15.79	2	<0.0001
Distance – tissue – Developmental stage interaction			
Tail length	52.16	4	<0.0001
Tail DNA	475.90	4	<0.0001
Olive moment	28.16	4	<0.0001
Tail moment	5.94	4	>0.05
% severed cells	8.95	4	>0.05

2.II.b Protein carbonyls, and lipid peroxides

The relative levels of protein carbonyls and lipid peroxides in males, females, and nymphs of *A. thalassinus* collected from different sites were shown in Fig. 26 and 27. The insects collected from the polluted sites A, B and C (1, 3, and 6 km respectively) away from the fertilizer factory showed significant increase in protein carbonyls and lipid peroxide level compared to the control site insects (Fig. 26 and 27). This increase was significantly correlated with the distance of the sites. A strong negative correlation among lipid peroxide content in the tissues and distance from factory was found (Table 12). In polluted sites, the variations in contents of carbonyls in respect to control site was higher in gut, followed by muscles and brain, respectively in males. In females, it was higher in gut followed by brain, while in nymphs, it was high in thoracic muscles followed by brain, and gut tissues (Table 12). Concentrations of carbonyls in insects' tissues tended to be high in nymphs, followed by females then in males.

The lipid peroxides concentrations in tissues of males, females, and nymphs from all sites differed significantly. In all tissues, the median values in insects from sites located along gradient of pollution were always significantly higher in relation to the control insects (Fig. 27). In males collected from site A, the median values in brain, thoracic muscles, and gut were 15.4, 23.5, and 11.5 times higher than control values, respectively. The tendency was always observed in case of males. Median concentrations of lipid peroxides in males' brain, thoracic muscles, and gut were 6.6, 7.3, and 4.0 times higher, respectively than in insects from the control group. The lipid peroxides concentrations were the highest in insects collected from site A. The lowest values were recorded mainly in individuals collected from site

C (Fig. 27). The increase of lipid peroxides level in female *A. thalassinus* which collected from site A was significantly higher in gut followed by thoracic muscles then brain. There was no significant difference among females' brain and thoracic muscles. In nymphs collected from site A, the median values in brain, thoracic muscles, and gut were 25.2, 11.4, and 15.3 times higher than control values, respectively. The lipid peroxides concentrations were the highest in insects collected from site A, while the lowest values were measured mainly in individuals from site C. Lipid peroxides levels in nymphs was the highest then in females, followed by males almost in all analyzed tissue. The interaction analysis using Generalized Estimating Equation (GEE) showed significant influence of distance, sex, and tissues on level of ROS, and macromolecules damage in males, females, and nymphs of *A. thalassinus* (Table 13). Also, interaction among distance, developmental stage, and tissues had a significant effect on level of ROS, and macromolecules damage (protein carbonyls, and lipid peroxides) (Table 14).

A cluster analysis using Ward's method revealed slightly dissimilar patterns for males, females and nymphs, however, the general tendency was similar (Fig 28-30). The levels of macromolecules damage (DNA strand breaks, protein carbonyls, and lipid peroxides) were highly similar in brain of males, and females collected from sites B and C. The level of macromolecules damage in brain tissue of males, females and nymphs from site A was also linked with those from sites B and C. Brain tissue of males from control site created a separate cluster. In all tissues of females, site A created a separate cluster. Finally, in gut tissue of nymphs from site A had a separate cluster. The distance between clusters of control site and polluted sites were relatively high.

Chi-square (x^2) test using cross tabulation were performed to examine independency among air pollutants of AZFC stack emission (SO_2, and TSP), pollutants in soil, plant, and water (heavy metals (Cu, Zn, Pb, and Cd), PO_4^{3-}, and SO_4^{2-}); and ROS (concentration of H_2O_2, and $O_2^{·-}$ production rate), in brain, thoracic muscles, and gut of insect samples. The result showed that there was a positive correlation (weak to strong) among air, soil, plant, and water pollutants; and ROS in gut tissue of male, female, and nymph insects (0.16, 0.64, and 0.42) ($p <$ 0.05) (Table 15).

The result showed that there was a strong positive correlation equal to 0.7 among ROS and macromolecules damage in thoracic muscles of males, and females insect. The ROS concentration had a synergistic effect on macromolecules damage ($p <$ 0.05) (Table 16).

Also, x^2 examined the relationship among air pollutants of AZFC stack emission (SO_2, and TSP), pollutants in soil, plant, and water (heavy metals (Cu, Zn, Pb, and Cd), PO_4^{3-}, and SO_4^{2-}); and macromolecules damage (DNA strand breaks, protein carbonyls amount, and lipid peroxides concentration) in brain, thoracic muscles, and gut of male, female, and nymph insect samples. The result showed that significant level of x^2 was lower than 0.05, and there was almost a positive correlation (weak to strong) among pollutants concentration in air, soil, plant, and water (Table 17). Also, the result showed that the combined effect of pollutants concentration in air, soil, plant, and water had a positive strong correlation (0.95, and 0.98) with protein carbonyls amount in brain tissue of male, and nymph, respectively.

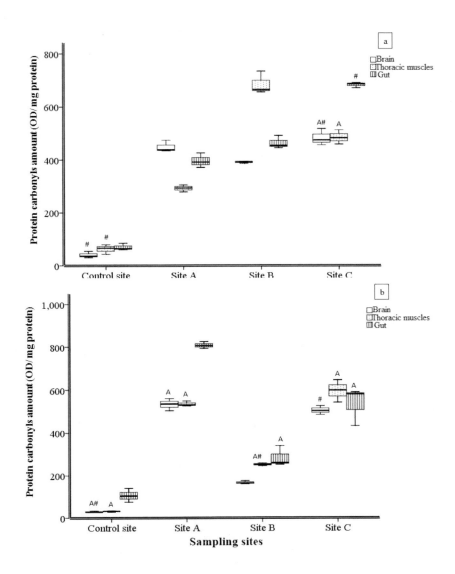

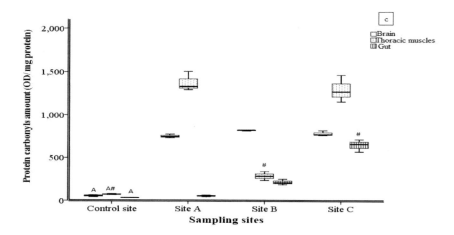

Fig. 26. The effect of the distance away from the fertilizer factory with respect to the control site on the amount of protein carbonyls (OD/ mg protein) in the brain, thoracic muscles and gut homogenates of *Aiolopus thalassinus* males (a), females (b), and 5ᵗʰ instar (c).

Median values marked with different small letters are significantly different among control and polluted sites located at different distances from the fertilizer factory. Median values marked with different capital letters are significantly different among tissues (Kruskal-Wallis revealed, $p < 0.05$). # denote no significant differences among males, females, and nymph in each case separately (Kruskal-Wallis revealed, $p < 0.05$).

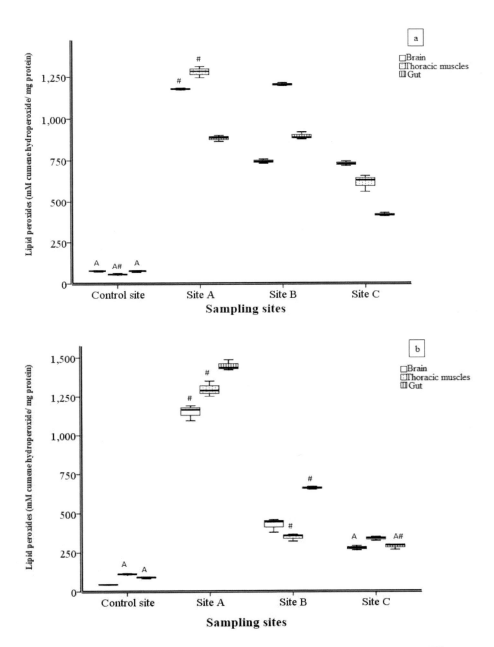

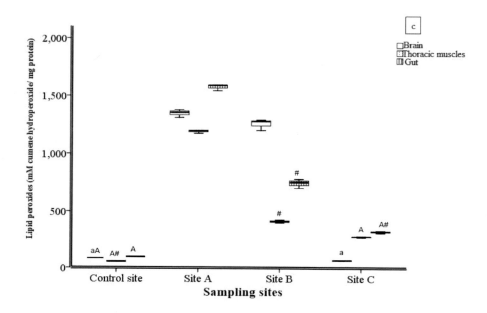

Fig.27. The effect of the distance away from the fertilizer factory with respect to the control site on the concentration of lipid peroxides (mM cumene hydroperoxide/ mg protein) in the brain, thoracic muscles and gut homogenates of *Aiolopus thalassinus* males (a), females (b), and 5th instar (c).

Median values marked with different small letters are significantly different among control and polluted sites located at different distances from the fertilizer factory. Median values marked with different capital letters are significantly different among tissues (Kruskal-Wallis revealed, $p < 0.05$). # denote no significant differences among males, females, and nymph in each case separately (Kruskal-Wallis revealed, $p < 0.05$).

Table 12. Pearson's correlation coefficient among macromolecules (proteins, and lipids) damage (protein carbonyls, and lipid peroxides) in brain, thoracic muscles, and gut homogenates of males, females, and 5th instar of *Aiolopus thalassinus* and the distance from fertilizer factory.

Damage	Sex	Brain Regression analysis	r	Thoracic muscles Regression analysis	r	Gut Regression analysis	r
Protein carbonyls	Male	$y= 64.51x^2 - 239.52x + 613$	+0.33	$y=-277.01x^2 + 1202x - 630$	+0.48	$y= 85.5x^2 - 194.50x + 5.21$	+0.94**
	Female	$y= 355x^2 - 1436x + 1615$	-0.07	$y= 312x^2 - 1214x + 1431$	+0.16	$y= 432x^2 - 1841.20x + 2214$	-0.51
	Nymph	$y= -122.81x^2 + 502.2x + 365$	+0.41	$y= 1.145x^2 - 4087x + 4401$	-0.06	$y= 149x^2 - 293x + 198$	+0.95**
Lipid peroxides	Male	$y= 62.51x^2 + 262.5x + 545$	-0.87**	$y= 1014x^2 - 4087x + 4401$	-0.90**	$y= 149x^2 - 293x + 198$	-0.85**
	Female	$y= 247x^2 - 1547x + 2434$	-0.93**	$y= 458x^2 - 2308x + 3139$	-0.86**	$y= 205x^2 - 1391x + 2621$	-0.97**
	Nymph	$y= 276x^2 - 1547x + 2414$	-0.95**	$y= 4589x^2 - 2318x + 3139$	-0.86**	$y= 204x^2 - 1393x + 2652$	-0.97**

* significant at $p<0.05$; ** significant at $p<0.001$.

81

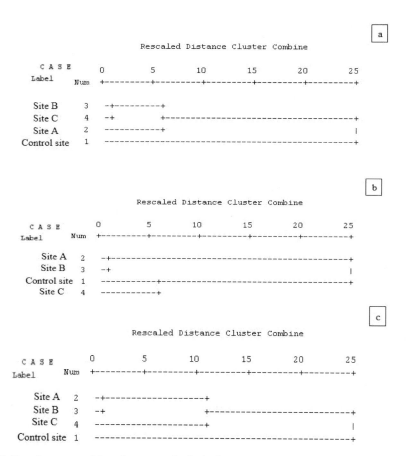

Fig. 28. Dendrogram of the cluster analysis (using Ward's Method) applied for biomarkers (DNA damage analysis, protein carbonyls amount, and lipid peroxides concentration) in brain (a), thoracic muscles (b), and gut (c) cells, and homogenates for (DNA, and protein carbonyls & lipid peroxides respectively) of males of *Aiolopus thalassinus*, which were collected at control site and polluted sites (A-C) located at different distances from fertilizer factory.

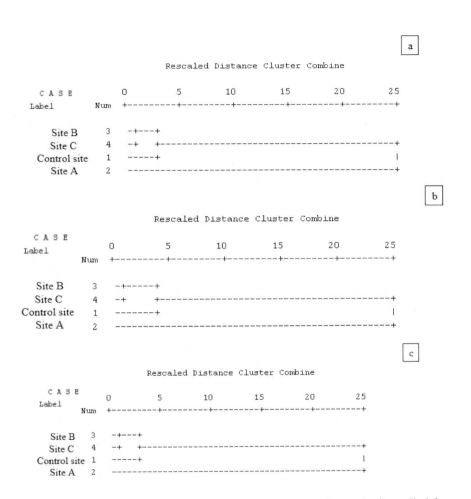

Fig. 29. Dendrogram of the cluster analysis (using Ward's Method) applied for biomarkers (DNA damage analysis, protein carbonyls amount, and lipid peroxides concentration) in brain (a), thoracic muscles (b), and gut (c) cells, and homogenates for (DNA, and protein carbonyls & lipid peroxides respectively) of females of *Aiolopus thalassinus*, which were collected at control site and polluted sites (A-C) located at different distances from fertilizer factory.

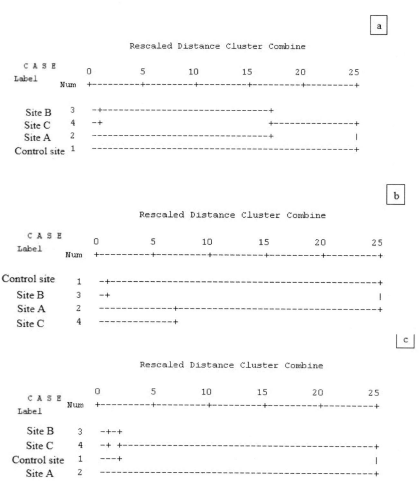

Fig. 30. Dendrogram of the cluster analysis (using Ward's Method) applied for biomarkers (DNA damage analysis, protein carbonyls amount, and lipid peroxides concentration) in brain (a), thoracic muscles (b), and gut (c) cells, and homogenates for (DNA, and protein carbonyls & lipid peroxides respectively) of 5th instar of *Aiolopus thalassinus*, which were collected at control site and polluted sites (A-C) located at different distances from the fertilizer factory.

Table 13. Generalized Estimating Equation to analyze the interactions among the distance from fertilizer factory, types of tissues, sex on ROS (H_2O_2 concentration and O_2^- production rate) and macromolecules damage (protein carbonyls amount and lipid peroxides concentration) in brain, thoracic muscles, and gut homogenates of males, and females of *Aiolopus thalassinus* collected at different sites located at various distances from the fertilizer factory.

Source	Chi-square (χ^2)	df	p value
Distance – sex interaction			
H_2O_2	102	2	<0.0001
O_2^-	3431	2	<0.0001
Protein carbonyls	1495	2	<0.0001
Lipid peroxides	1539	2	<0.0001
Distance – tissue interaction			
H_2O_2	201	4	<0.0001
O_2^-	21911	4	<0.0001
Protein carbonyls	418	4	<0.0001
Lipid peroxides	645	4	<0.0001
Sex – tissue interaction			
H_2O_2	294	2	<0.0001
O_2^-	3367	2	<0.0001
Protein carbonyls	12	2	<0.05
Lipid peroxides	1087	2	<0.0001
Distance – tissue – sex interaction			
H_2O_2	92	4	<0.0001
O_2^-	5088	4	<0.0001
Protein carbonyls	251	4	<0.0001
Lipid peroxides	525	4	<0.0001

Table 14. Generalized Estimating Equation to analyze the interactions among the distance from fertilizer factory, types of tissues, developmental stage (adult and nymph) on ROS (H_2O_2 concentration and O_2^- production rate) and macromolecules damage (protein carbonyls amount and lipid peroxides concentration) in brain, thoracic muscles, and gut homogenates of males, females, and 5th instar of *Aiolopus thalassinus* collected at different sites located at various distances from the fertilizer factory.

Source	Chi-square (χ^2)	df	p value
Distance – Developmental stage interaction			
H_2O_2	35	2	<0.0001
O_2^-	131	2	<0.0001
Protein carbonyls	42	2	<0.0001
Lipid peroxides	25	2	<0.0001
Distance – tissue interaction			
H_2O_2	107	4	<0.0001
O_2^-	114	4	<0.0001
Protein carbonyls	99	4	<0.0001
Lipid peroxides	25	4	<0.0001
Developmental stage – tissue interaction			
H_2O_2	22	2	<0.0001
O_2^-	38	2	<0.0001
Protein carbonyls	241	2	<0.0001
Lipid peroxides	12	2	<0.0001
Distance – tissue – Developmental stage interaction			
H_2O_2	27	4	<0.0001
O_2^-	107	4	<0.0001
Protein carbonyls	160	4	<0.0001
Lipid peroxides	4	4	<0.0001

Table 15. Chi-square (x^2) using cross tabulation was performed to examine independency among air, soil, plant, and water pollutants concentration (heavy metals (Cu, Zn, Pb, and Cd), and PO_4^{3-}, and SO_4^{2-}) and reactive oxygen species amount (ROS) (production rate of superoxide anion ($O_2^{\bullet}$) and concentration of hydrogen peroxide (H_2O_2)) in brain, thoracic muscles, and gut homogenates of males, females, and 5th instar of *Aiolopus thalassinus*. Samples collected at polluted sites (A-C) located at different distances from the fertilizer factory.

	Tissue	X^2	LR	*df*	LLA	*df*	R	r	T	*p* value
Male	Brain	396	133	378	0.57	1	-0.15	0.24	11.7	< 0.0001
	Thoracic muscle	402	133	378	0.01	1	-0.02	0.41	10.7	< 0.0001
	Gut	432	138	414	0.64	1	0.16	0.61	14.7	< 0.0001
Female	Brain	372	129	360	1.91	1	0.28	0.37	9.3	< 0.0001
	Thoracic muscle	384	130	360	0.001	1	0.006	0.75	14.7	< 0.0001
	Gut	432	138	414	9.5	1	0.64	0.72	14.7	< 0.0001
Nymph	Brain	402	133	378	1.4	1	0.24	0.57	10.7	< 0.0001
	Thoracic muscle	432	138	414	0.05	1	-0.04	0.60	14.7	< 0.0001
	Gut	432	138	414	4.2	1	0.42	0.71	14.7	< 0.0001

X^2, LR, LLA, R, r, and T represent of Person's chi-square, Likelihood Ratio, Linear by Linear Association, Person's R, correlation, and T of Lambda respectively. x^2 revealed was made for each tissue (brain, thoracic muscles, and gut), each sex and developmental stage (male, female, and nymph) separately.

Table 16. Chi-square (x^2) using cross tabulation was performed to examine independency among reactive oxygen species amount (ROS) (production rate of superoxide anion (O_2^-) and concentration of hydrogen peroxide (H_2O_2)) and macromolecules damage (DNA strand breaks, protein carbonyls amount, and lipid peroxides concentration) in brain, thoracic muscles, and gut homogenates of males, females, and 5th instar of *Aiolopus thalassinus*.

	Tissue	X^2	LR	*df*	LLA	*df*	R	r	T	*p* value
Male	Brain	504	147	483	3.7	1	0.40	0.73	17.6	< 0.0001
	Thoracic muscle	504	147	483	13.3	1	0.76	0.349	17.6	< 0.0001
	Gut	552	152	529	8.2	1	0.59	0.29	33.9	< 0.0001
Female	Brain	456	140	440	2.31	1	0.32	0.22	11.43	< 0.0001
	Thoracic muscle	480	144	460	13.4	1	0.76	0.23	16.47	< 0.0001
	Gut	552	152	529	2.89	1	0.35	0.27	33.9	< 0.0001
Nymph	Brain	480	144	462	0.007	1	-0.017	0.06	13.69	< 0.0001
	Thoracic muscle	552	152	529	2.03	1	0.29	0.33	33.9	< 0.0001
	Gut	582	149	506	8.2	1	0.60	0.42	19.05	< 0.0001

X^2, LR, LLA, R, r, and T represent of Person's chi-square, Likelihood Ratio, Linear by Linear Association, Person's R, correlation, and T of Lambda respectively. x^2 revealed was made for each tissue (brain, thoracic muscles, and gut), each sex and developmental stage (male, female, and nymph) separately.

Table 17. Chi-square (x^2) revealed using cross tabulation to revealed independency among air (SO$_2$, and TSP), soil, plant, and water pollutants, each source separately and combined, concentration (heavy metals (Cu, Zn, Pb, and Cd), and PO$_4^{3-}$, and SO$_4^{2-}$) and macromolecules damage (DNA strand breaks) in brain, thoracic muscles, and gut homogenates of males, females, and 5th instar of *Aiolopus thalassinus*. Samples collected at polluted sites (A-C) located at different distances from fertilizer factory.

MD		Pollutant source	Air	Soil	Plant	Water	Pollutants*	*p* value
		Tissue	R	R	R	R	R	
DNA	Male	Brain	0.93	0.31	0.34	0.16	0.14	< 0.05
		Thoracic muscle	0.91	0.22	0.44	0.27	0.16	< 0.05
		Gut	0.92	0.35	0.34	0.13	0.12	< 0.05
	Female	Brain	0.88	0.26	0.52	0.31	0.16	< 0.05
		Thoracic muscle	0.89	0.33	0.40	0.19	0.19	< 0.05
		Gut	0.97	0.25	0.52	0.31	0.17	< 0.05
	Nymph	Brain	0.94	0.22	0.38	0.23	0.16	< 0.05
		Thoracic muscle	0.96	0.22	0.54	0.33	0.18	< 0.05
		Gut	0.98	0.28	0.49	0.27	0.15	< 0.05
Protein	Male	Brain	0.97	0.51	0.13	0.49	0.95	< 0.05
		Thoracic muscle	0.96	0.55	0.49	0.66	0.83	< 0.05
		Gut	0.96	0.24	0.04	0.27	0.89	< 0.05
	Female	Brain	0.97	0.23	0.23	0.13	0.68	< 0.05
		Thoracic muscle	0.97	0.25	0.17	0.16	0.77	< 0.05
		Gut	0.98	0.43	0.004	0.31	0.62	< 0.05
	Nymph	Brain	0.97	0.65	0.32	0.64	0.98	< 0.05
		Thoracic muscle	0.97	0.17	0.26	0.05	0.61	< 0.05
		Gut	0.86	0.26	0.37	0.22	0.53	< 0.05
Lipid	Male	Brain	0.97	0.73	0.29	0.63	0.80	< 0.05
		Thoracic muscle	0.97	0.93	0.61	0.89	0.80	< 0.05
		Gut	0.97	0.94	0.66	0.90	0.78	< 0.05
	Female	Brain	0.97	0.72	0.30	0.57	0.52	< 0.05
		Thoracic muscle	0.97	0.64	0.19	0.48	0.48	< 0.05
		Gut	0.97	0.80	0.41	0.67	0.53	< 0.05
	Nymph	Brain	0.97	0.71	0.30	0.56	0.50	< 0.05
		Thoracic muscle	0.97	0.61	0.17	0.45	0.41	< 0.05
		Gut	0.97	0.79	0.41	0.65	0.49	< 0.05
**Combined	Male	Brain	0.97	0.39	0.28	0.36	0.06	< 0.05
		Thoracic muscle	0.96	0.37	0.27	0.34	0.06	< 0.05
		Gut	0.96	0.41	0.30	0.38	0.05	< 0.05
	Female	Brain	0.97	0.34	0.24	0.31	0.04	< 0.05
		Thoracic muscle	0.97	0.33	0.24	0.31	0.05	< 0.05
		Gut	0.98	0.35	0.25	0.32	0.001	< 0.05
	Nymph	Brain	0.97	0.38	0.28	0.36	0.03	< 0.05
		Thoracic muscle	0.97	0.34	0.25	0.32	0.01	< 0.05
		Gut	0.86	0.30	0.21	0.28	0.08	< 0.05

R, and MD represent Person's correlation, and macromolecules damage (DNA strand breaks, protein carbonyls amounts, and lipid peroxides concentration), respectively. * pollutants mean the combined effect of air, soil, plant, and water. ** combined means combined effect of pollutants on DNA, protein, and lipid. x^2 revealed was made for each tissue (brain, thoracic muscles, and gut), each sex and developmental stage (male, female, and nymph) separately.

3 oxidative stress responses

3.I Non- enzymatic antioxidants assay

3.I.a Glutathione reduced (GSH)

The relative levels of non-enzymatic antioxidants (GSH concentration) in males, females, and nymphs of *A. thalassinus* are shown at Fig. 31. The concentration of GSH in thoracic muscles tissue of males, and females, and brain tissue of nymphs collected at site A was significantly lower than in gut tissue. There was no significant difference among brain, and gut tissues in males, and nymphs. The median values of GSH concentration in all tissues of males, females, and nymphs at site A were significantly different from control site ($p<$ 0.05) except for thoracic muscles, brain, gut tissues of males, females, and nymphs, respectively ($p>$ 0.05) (Fig. 31).

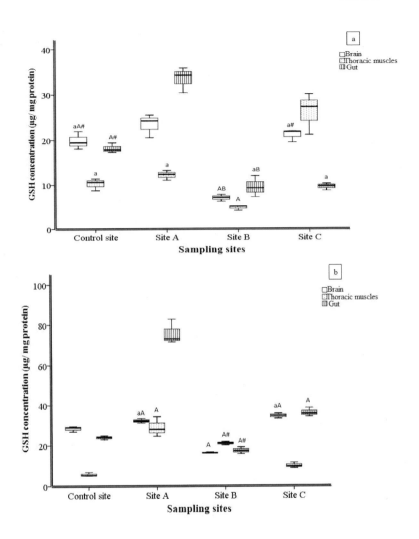

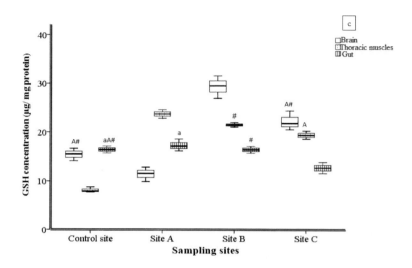

Fig. 31. Glutathione reduced (GSH) concentration, expressed as median, percentile deviation (P25 and P75 -boxes), and min-max values of GSH concentration obtained from brain, thoracic muscles, and gut homogenates of males (a), females (b), and nymph (c) of *Aiolopus thalassinus* collected at different distances from fertilizer factory.

Median values marked with different small letters are significantly different among control and polluted sites (A-C) located at different distances from the fertilizer factory. Median values marked with different capital letters are significantly different among tissues (Kruskal-Wallis revealed, $p < 0.05$). # denote no significant differences among males, females, and nymph in each case separately (Kruskal-Wallis revealed, $p < 0.05$).

3.II Enzymatic antioxidants assays

The relative levels of antioxidant enzymes in males, females, and nymphs of *A. thalassinus* are shown in Fig 32-38. The activity of superoxide dismutase (SOD) in gut of males collected at site A was significantly lower than in thoracic muscles. While, the activity of SOD in brain of females and nymphs from the same site was significantly higher than in thoracic muscles and gut. (Fig. 32).

The level of CAT activity in males, females, and nymphs from site A (nearby the factory) revealed a significant decrease compared to individuals from the control site (Fig. 33). In case of females from polluted sites, strong inhibition of CAT activity was observed. The lowest CAT activity value in brain tissue of males, and nymphs of *A. thalassinus* occurs in case of individuals collected from sites A and C, respectively. Compared to control insects, the median values in brain, thoracic muscles and gut of males from site (A) were, respectively, 90.20%, 68.41%, and 84.16% lower.

The present work showed that, there was no clear tendency to describe a relation among pollution and APOX activity, especially in case of males (Fig. 34). In males, the lowest APOX activity was observed in individuals form site A (in brain), B (in gut), and C (in brain). In females, the lowest APOX activity values were noted in specimen collected from sited located 6, and 1 km from the source of contamination (AZFC). Interestingly, in females the highest APOX values were observed in thoracic muscle tissues. In nymphs, there was a significantly difference in almost all tissues from site A-C with respect to control site ($p < 0.05$).

The results show that distance from the factory influenced PPO activity in *A. thalassinus* in case of each tissue (Fig. 35). The highest values were observed in muscles, both in males and females; and in gut of nymphs from all analyzed sites. PPO activity in males, and nymphs collected from site C and A, respectively achieved the highest values for all tissues. Also in males, females, and nymphs of *A. thalassinus* collected at site A, PPO activity was significantly increased with respect to the insects collected from the control site.

The results showed that in control site, there was no significantly difference of POX activity among brain, and thoracic muscles tissue in males; and among all tissues in females, and nymphs ($p> 0.05$). Also, there was no significant difference among females, and nymphs in brain tissue of insect at site A ($p> 0.05$) (Fig. 36). The POX activity in brain, thoracic muscles, and gut tissues of males, females, and nymphs was significantly higher in specimens collected from site A than control site ($p < 0.05$).

The activity of glutathione reductase (GR) in thoracic muscles of males, and females collected at site A was significantly lower than in gut. Although, the activity of GR in nymphs from the same site was significantly lower in gut than thoracic muscles. The highest value of GR activity occurred in thoracic muscles tissues of males, and females at site C, and in gut tissue of nymph tissues at site A (Fig. 37).

In the tissues of males, females, and nymphs from site A (nearby the factory) the level of glutathione-s-transferase (G-S-T) activity revealed a significant increasing effect compared to individuals from the control site (Fig. 38). The highest G-S-T activity value in gut tissue of males, females, and nymphs of *A. thalassinus* occurs in individuals collected from site C, B, and A respectively.

The concentration of non-enzymatic response (GSH), and activity of all antioxidant enzymes (SOD, CAT, APOX, PPO, POX, GR, and G-S-T) was significantly higher in nymphs than in females and males in all analyzed tissues. Distance of sampling sites from the fertilizer factory and antioxidants response (non-enzymatic and enzymatic antioxidants) in *A. thalanthsis* were negatively or positively correlated which depend on the role of each antioxidant (Table 18).

The correlation of activity of SOD and CAT in all tissues in male, female, and nymph was significantly positive correlation ($P<$ 0.05) (Table 19). While, the correlation among activity of CAT and APOX, and POX was negatively significant correlation ($P<$ 0.05) (Table 19).

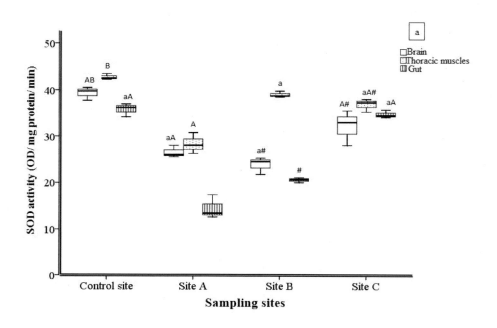

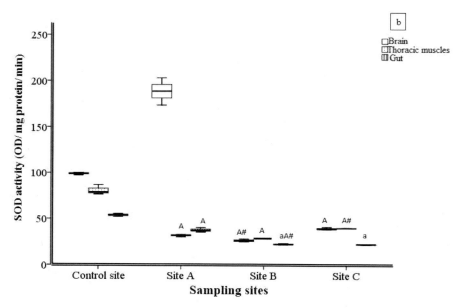

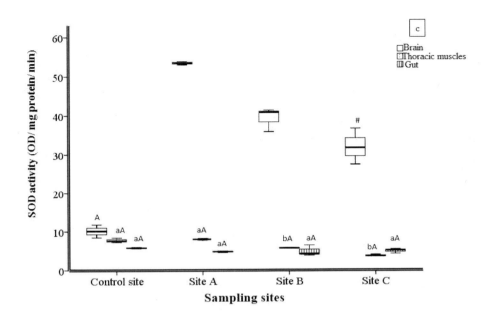

Fig. 32. Activity of superoxide dismutase (SOD), expressed as median, percentile deviation (P25 and P75 -boxes), and min-max values of SOD activity obtained from brain, thoracic muscles, and gut homogenates of males (a), females (b), and nymph (c) of *Aiolopus thalassinus* collected at different distances from Fertilizer factory. median values marked with different small letters are significantly different among control and polluted sites (A-C) located at different distances from the fertilizer factory.

Median values marked with different capital letters are significantly different among tissues (Kruskal-Wallis revealed, $p < 0.05$). # denote no significant differences among males, females, and nymph in each case separately (Kruskal-Wallis revealed, $p < 0.05$).

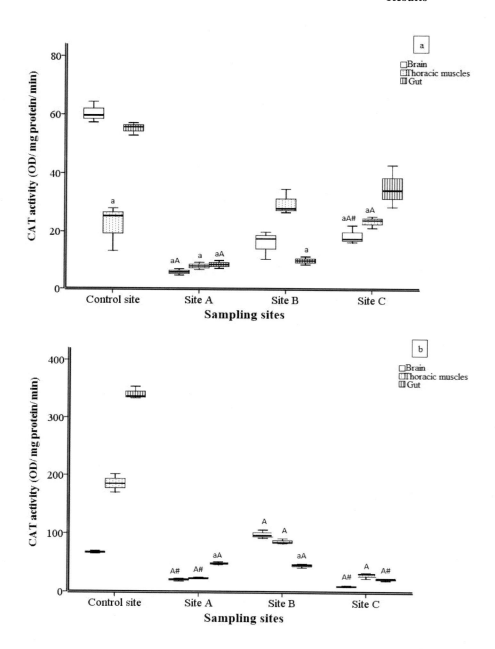

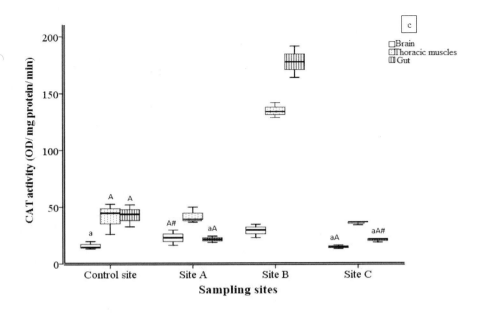

Fig. 33. Activity of catalase (CAT), expressed as median, percentile deviation (P25 and P75 -boxes), and min-max values of CAT activity obtained from brain, thoracic muscles, and gut homogenates of males (a), females (b), and nymph (c) of *Aiolopus thalassinus* collected at different distances from fertilizer factory.

median values marked with different small letters are significantly different among control and polluted sites (A-C) located at different distances from the fertilizer factory. Median values marked with different capital letters are significantly different among tissues (Kruskal-Wallis revealed, $p < 0.05$). # denote no significant differences among males, females, and nymph in each case separately (Kruskal-Wallis revealed, $p < 0.05$).

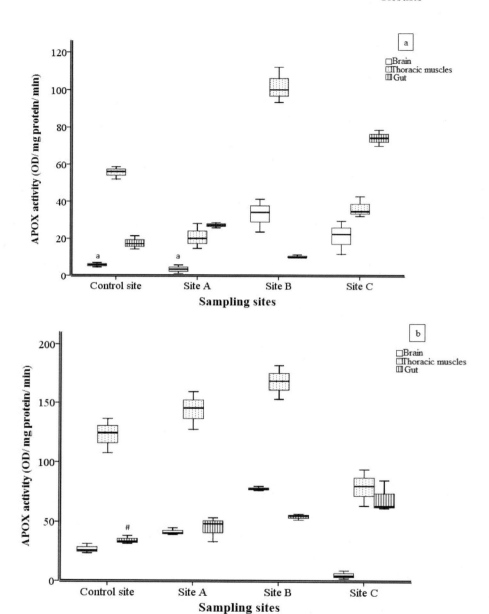

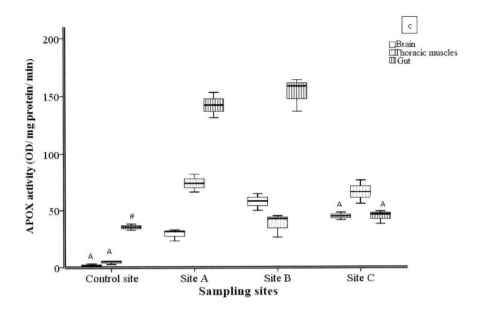

Fig. 34. Activity of ascorbate peroxidase (APOX), expressed as median, percentile deviation (P25 and P75 -boxes), and min-max values of APOX activity obtained from brain, thoracic muscles, and gut homogenates of males (a), females (b), and nymph (c) of *Aiolopus thalassinus* collected at different distances from the fertilizer factory.
Median values marked with different small letters are significantly different among control and polluted sites (A-C) located at different distances from the fertilizer factory. Median values marked with different capital letters are significantly different among tissues (Kruskal-Wallis revealed, $p < 0.05$). # denote no significant differences among males, females, and nymph in each case separately (Kruskal-Wallis revealed, $p < 0.05$).

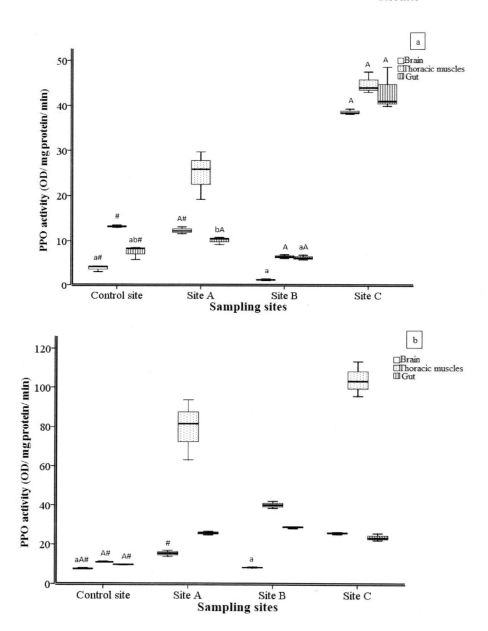

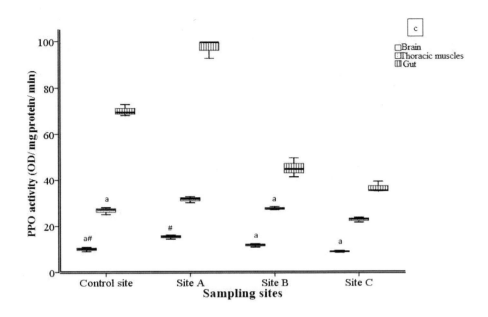

Fig. 35. Activity of polyphenol oxidase (PPO), expressed as median, percentile deviation (P25 and P75 -boxes), and min-max values of PPO activity obtained from brain, thoracic muscles, and gut homogenates of males (a), females (b), and nymph (c) of *Aiolopus thalassinus* collected at different distances from fertilizer factory.

Median values marked with different small letters are significantly different among control and polluted sites (A-C) located at different distances from the fertilizer factory. Median values marked with different capital letters are significantly different among tissues (Kruskal-Wallis revealed, $p < 0.05$). # denote no significant differences among males, females, and nymph in each case separately (Kruskal-Wallis revealed, $p < 0.05$).

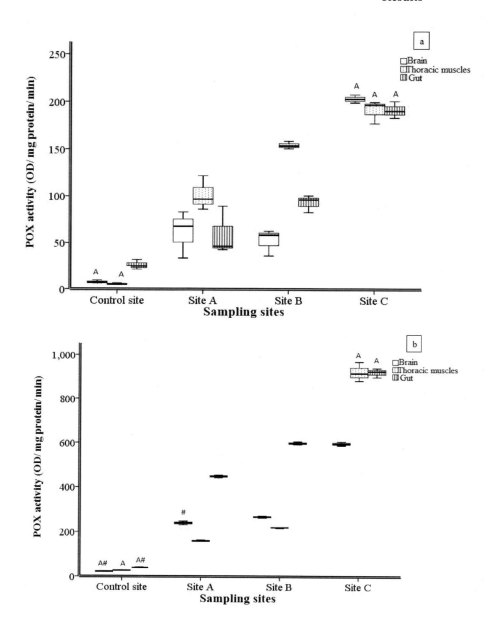

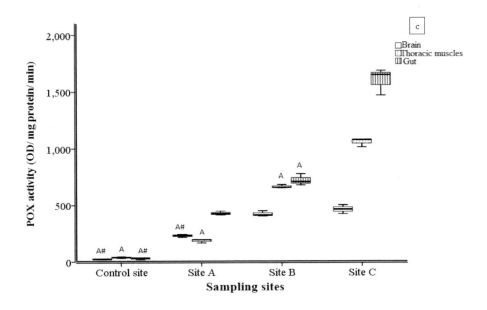

Fig. 36. Activity of peroxidase (POX), expressed as median, percentile deviation (P25 and P75 -boxes), and min-max values of POX activity obtained from brain, thoracic muscles, and gut homogenates of males (a), females (b), and nymph (c) of *Aiolopus thalassinus* collected at different distances from fertilizer factory.

Median values marked with different small letters are significantly different among control and polluted sites (A-C) located at different distances from the fertilizer factory. Median values marked with different capital letters are significantly different among tissues (Kruskal-Wallis revealed, $p < 0.05$). # denote no significant differences among males, females, and nymph in each case separately (Kruskal-Wallis revealed, $p < 0.05$).

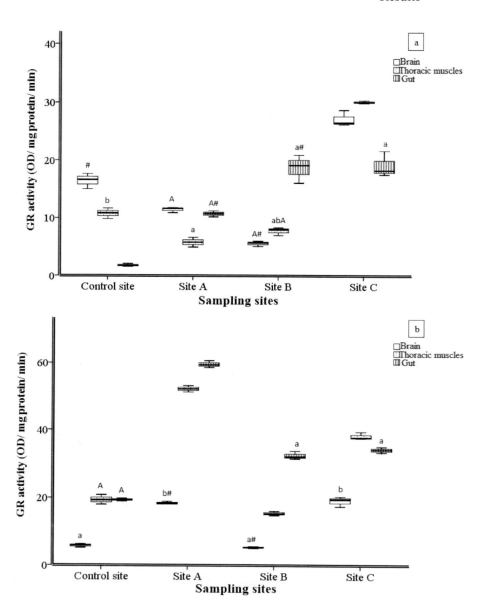

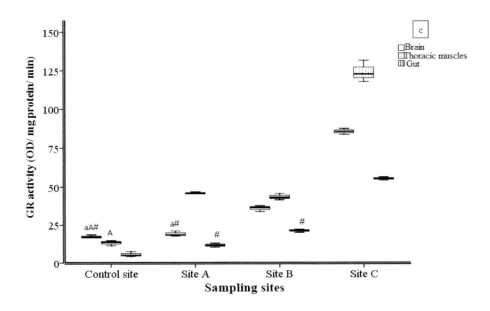

Fig. 37. Activity of glutathione reductase (GR), expressed as median, percentile deviation (P25 and P75 -boxes), and min-max values of GR activity obtained from brain, thoracic muscles, and gut homogenates of males (a), females (b), and nymph (c) of *Aiolopus thalassinus* collected at different distances from fertilizer factory.

Median values marked with different small letters are significantly different among control and polluted sites (A-C) located at different distances from the fertilizer factory. Median values marked with different capital letters are significantly different among tissues (Kruskal-Wallis revealed, $p < 0.05$). # denote no significant differences among males, females, and nymph in each case separately (Kruskal-Wallis revealed, $p < 0.05$).

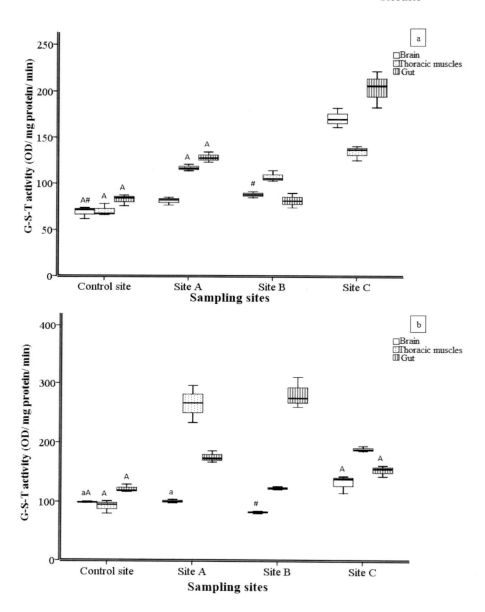

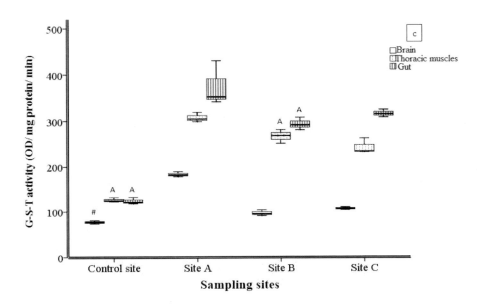

Fig. 38. Activity of glutathione-s-transferase (G-S-T), expressed as median, percentile deviation (P25 and P75 -boxes), and min-max values of G-S-T activity obtained from brain, thoracic muscles, and gut homogenates of males (a), females (b), and nymph (c) of *Aiolopus thalassinus* collected at different distances from fertilizer factory.

Median values marked with different small letters are significantly different among control and polluted sites (A-C) located at different distances from the fertilizer factory. Median values marked with different capital letters are significantly different among tissues (Kruskal-Wallis revealed, $p < 0.05$). # denote no significant differences among males, females, and nymph in each case separately (Kruskal-Wallis revealed, $p < 0.05$).

Table 18. Pearson's correlation coefficient among oxidative stress response (non-enzymatic, and enzymatic) response (GSH, SOD, CAT, APOX, PPO, POX, GR, and G-S-T) in brain, thoracic muscles, and gut homogenates of males, females, and 5th instar of *Aiolopus thalassinus* and the distance from fertilizer factory.

Oxidative stress response	Sex	Brain — Regression analysis	Brain — r	Thoracic muscles — Regression analysis	Thoracic muscles — r	Gut — Regression analysis	Gut — r
GSH	Male	$y= 2.6x^2 - 19.2x + 40.8$	-0.13	$y= 2.1x^2 - 12.3x + 22.4$	0.62	$y= 2.5x^2 - 22.6x + 54.3$	-0.85**
	Female	$y= 2.8x^2 - 19.3x + 48.6$	0.11	$y= -0.06x^2 - 3.2x + 31.4$	-0.95**	$y= 6.8x^2 - 55.2x + 121.5$	-0.65
	Nymph	$y= -2.3x^2 - 18.1x - 4.3$	0.58	$y= -0.08x^2 - 1.4x + 24.9$	-0.94**	$y= -0.17x^2 + 0.35x + 16.9$	-0.86**
SOD	Male	$y= 0.8x^2 - 4.2x + 29.4$	0.57	$y= -1.1x^2 + 9.2x + 19.8$	0.72*	$y= 0.2x^2 + 2.5x + 10.2$	0.96**
	Female	$y= 17x^2 - 149x + 320$	-0.82**	$y= 1.1x^2 - 6.5x + 37.4$	0.70*	$y= 1.5x^2 - 13.5x + 49$	-0.99**
	Nymph	$y= 0.7x^2 - 9.3x + 61.6$	0.96**	$y= 0.1x^2 - 2.1x + 10$	-0.86**	$y= 0.06x^2 - 0.2x + 4.2$	0.093
CAT	Male	$y= -1.7x^2 + 12.9x - 6.2$	0.84**	$y= -2.2x^2 + 19.1x - 8.8$	0.67*	$y= 1.4x^2 - 4.6x + 11.2$	0.85**
	Female	$y= -13.4x^2 + 91.6x - 57.2$	-0.11	$y= -9.8x^2 + 70.2x - 37.4$	0.07	$y= -1.3x^2 + 4.3x + 45$	-0.90**
	Nymph	$y= -1.7x^2 + 10.3x + 14.4$	-0.46	$y= -16.0x^2 + 111.6x - 56.6$	-0.05	$y= -26.1x^2 + 183.1x - 136$	-0.005

Results

Continue table 18. Pearson's correlation coefficient among oxidative stress response (non-enzymatic, and enzymatic) response (GSH, SOD, CAT, APOX, PPO, POX, GR, and G-S-T) in brain, thoracic muscles, and gut homogenates of males, females, and 5th instar of *Aiolopus thalassinus* and the distance from fertilizer factory.

Oxidative stress response	Tissue Sex	Brain Regression analysis	Thoracic muscles r	Gut Regression analysis	Oxidative stress response	Tissue Sex	Brain Regression analysis
	Male	$y= -3.9x^2 + 31.1x - 24.2$	0.53	$y= -12.4x^2 + 89.6x - 57.2$	0.17	$y= 5.9x^2 - 32.3x + 53.4$	0.71*
APOX	Female	$y= -8.5x^2 + 52.7x - 4.2$	-0.49	$y= -8.1x^2 + 44.1x + 109$	-0.67*	$y= -0.1x^2 + 4.1x + 44$	0.78*
	Nymph	$y= -3.5x^2 + 27.7x + 6.8$	0.51	$y= 4.8x^2 - 35.2x + 104.4$	-0.18	$y= -9.1x^2 + 44.2x + 106.8$	-0.80**
	Male	$y= 3.5x^2 - 19.7x + 28.2$	0.68*	$y= 4.4x^2 - 27.2x + 47.8$	0.51	$y= 2.6x^2 - 12.6x + 20$	0.80**
PPO	Female	$y= 1.8x^2 - 10.8x + 24$	0.58	$y= 8.3x^2 - 53.7x + 126.4$	0.36	$y= -0.6x^2 + 4.0x + 21.6$	-0.36
	Nymph	$y= 0.2x^2 - 2.8x + 17.6$	-0.97**	$y= 0.1x^2 - 2.5x + 33.4$	-0.97**	$y= 4.9x^2 - 47.1x + 141.2$	-0.92**
	Male	$y= 10.5x^2 - 46.5x + 103$	0.82**	$y= -2.8x^2 + 39.8x + 59$	0.95**	$y= 1.2x^2 + 19.9x + 24.8$	0.93**
POX	Female	$y= 19.1x^2 - 62.2x + 282.3$	0.89**	$y= 40.2x^2 - 131.1x + 249.8$	0.89**	$y= 6.8x^2 - 47.1x + 393$	0.97**
	Nymph	$y= -14.4x^2 + 146.6x + 104.8$	0.91**	$y= -18.2x^2 + 303.8x - 87.6$	0.99**	$y= 33.8x^2 + 8.8x + 382.4$	0.95**

Continue table 18. Pearson's correlation coefficient among oxidative stress response (non-enzymatic, and enzymatic) response (GSH, SOD, CAT, APOX, PPO, POX, GR, and G-S-T) in brain, thoracic muscles, and gut homogenates of males, females, and 5th instar of *Aiolopus thalassinus* and the distance from fertilizer factory.

Oxidative stress response	Tissue	Brain	Thoracic muscles	Gut	Oxidative stress response	Tissue	Brain
	Sex	Regression analysis	r	Regression analysis		Sex	Regression analysis
GR	**Male**	$y= 3.0x^2 - 11.0x + 20$	**0.70***	$y= 1.1x^2 - 3.1x + 7$	**0.89****	$y= -0.9x^2 + 8.3x + 2.6$	**0.82****
	Female	$y= 2.2x^2 - 15.4x + 31.2$	**0.042**	$y= 5.1x^2 - 39.1x + 86$	**-0.37**	$y= 2.8x^2 - 24.8x + 81$	**-0.83****
	Nymph	$y= 1.4x^2 + 3.1x + 13.4$	**0.96****	$y= 5.6x^2 - 24.0x + 64.4$	**0.84****	$y= 1.3x^2 - 0.4x + 11$	**0.94****
G-S-T	**Male**	$y= 5.0x^2 - 17.6x + 94.6$	**0.89****	$y= 3.1x^2 - 18.1x + 131$	**0.57**	$y= 12.9x^2 - 74.7x + 188.8$	**0.60**
	Female	$y= 5.6x^2 - 31.4x + 125.8$	**0.59**	$y= 18.7x^2 - 146.9x + 395.2$	**-0.51**	$y= -18.1x^2 +123.5x + 67.6$	**-0.15**
	Nymph	$y= 9.4x^2 - 81.2x + 203.8$	**-0.80****	$y= 1.4x^2 - 24.2x + 327.8$	**-0.90****	$y= 7.7x^2 - 61.3x + 405.6$	**-0.57**

* significant at p<0.05; ** significant at p<0.001.

Table 19. Pearson's correlation coefficient among oxidative stress response (non-enzymatic, and enzymatic) response (GSH, SOD, CAT, APOX, PPO, POX, GR, and G-S-T) in brain, thoracic muscles, and gut homogenates of males, females, and 5th instar of *Aiolopus thalassinus*.

D	T	AOS	GSH	SOD	CAT	APOX	PPO	POX	GR	G-S-T
		GSH	1							
		SOD	0.41	1						
		CAT	0.07	0.83	1					
	B	APOX	-0.77	-0.35	-0.28	1				
		PPO	0.49	0.11	-0.29	0.04	1			
		POX	0.24	-0.09	-0.43	0.30	0.94	1		
		GR	0.58	0.57	0.19	-0.15	0.87	0.72	1	
		G-S-T	0.20	-0.02	-0.32	0.32	0.93	0.96	0.76	1
		GSH	1							
		SOD	-0.17	1						
		CAT	-0.12	0.69	1					
M	T	APOX	-0.58	0.57	0.74	1				
		PPO	0.91	-0.34	-0.24	-0.70	1			
		POX	0.43	-0.28	0.23	0.09	0.50	1		
		GR	0.88	0.17	0.25	-0.26	0.82	0.52	1	
		G-S-T	0.61	-0.59	-0.10	-0.29	0.67	0.87	0.50	1
		GSH	1							
		SOD	-0.55	1						
		CAT	-0.27	0.91	1					
	G	APOX	-0.25	0.43	0.20	1				
		PPO	-0.38	0.47	0.21	0.97	1			
		POX	-0.54	0.27	-0.09	0.84	0.87	1		
		GR	-0.47	-0.21	-0.55	0.43	0.48	0.78	1	
		G-S-T	-0.13	0.25	0.04	0.95	0.94	0.82	0.47	1
		GSH	1							
		SOD	0.45	1						
		CAT	-0.92	-0.41	1					
	B	APOX	-0.89	-0.09	0.77	1				
		PPO	0.71	-0.08	-0.85	-0.68	1			
F		POX	0.33	-0.41	-0.56	-0.37	0.89	1		
		GR	0.79	0.33	-0.94	-0.57	0.87	0.67	1	
		G-S-T	0.77	-0.08	-0.80	-0.85	0.87	0.67	0.72	1
		GSH	1							
		SOD	-0.74	1						
	T	CAT	-0.60	0.85	1					
		APOX	0.62	-0.27	0.15	1				
		PPO	0.30	-0.62	-0.90	-0.49	1			

AOS: Antioxidant system, D: Development, M: Male, F: Female, N: Nymph, T: Tissue, B: Brain, T: Thoracic muscles, G: Gut.

Continue table 19. Pearson's correlation coefficient among oxidative stress response (non-enzymatic, and enzymatic) response (GSH, SOD, CAT, APOX, PPO, POX, GR, and G-S-T) in brain, thoracic muscles, and gut homogenates of males, females, and 5th instar of *Aiolopus thalassinus*.

D	T	AOS	GSH	SOD	CAT	APOX	PPO	POX	GR	G-S-T
N	G	POX	-0.24	-0.33	-0.58	-0.70	0.78	1		
		GR	0.48	-0.38	-0.74	-0.26	0.73	0.28	1	
		G-S-T	0.69	-0.59	-0.83	-0.042	0.76	0.26	0.94	1
		GSH	1							
		SOD	0.09	1						
		CAT	-0.33	0.89	1					
		APOX	-0.05	-0.81	-0.68	1				
		PPO	0.25	-0.85	-0.94	0.53	1			
		POX	0.05	-0.93	-0.88	0.87	0.73	1		
		GR	0.89	-0.24	-0.64	0.09	0.62	0.28	1	
		G-S-T	-0.25	-0.62	-0.54	0.17	0.75	0.33	0.15	1
	B	GSH	1							
		SOD	-0.05	1						
		CAT	0.36	0.47	1					
		APOX	0.73	0.57	0.56	1				
		PPO	-0.43	0.75	0.51	0.03	1			
		POX	0.67	0.50	0.25	0.92	-0.13	1		
		GR	0.45	0.01	-0.32	0.52	-0.58	0.75	1	
		G-S-T	-0.53	0.84	-0.21	0.10	0.84	0.08	-0.18	1
	T	GSH	1							
		SOD	-0.24	1						
		CAT	0.30	-0.12	1					
		APOX	0.85	-0.31	-0.17	1				
		PPO	0.31	0.78	0.10	0.13	1			
		POX	0.42	-0.96	0.20	0.44	-0.67	1		
		GR	0.41	-0.86	-0.23	0.63	-0.61	0.88	1	
		G-S-T	0.97	-0.14	0.25	0.84	0.40	0.33	0.36	1
	G	GSH	1							
		SOD	0.04	1						
		CAT	0.27	-0.06	1					
		APOX	0.51	-0.34	0.56	1				
		PPO	0.68	-0.01	-0.40	0.27	1			
		POX	-0.81	-0.29	-0.07	-0.13	-0.69	1		
		GR	-0.84	-0.23	-0.15	-0.29	-0.72	0.98	1	
		G-S-T	-0.04	-0.54	-0.03	0.61	0.11	0.50	0.38	1

AOS: Antioxidant system, D: Development, M: Male, F: Female, N: Nymph, T: Tissue, B: Brain, T: Thoracic muscles, G: Gut.

A cluster analysis using Ward's method revealed slightly dissimilar patterns for males, females and nymphs, however, the similar general tendency (Fig 39-41). The level of non-enzymatic (GSH concentration), and enzymatic response (SOD, CAT, APOX, PPO, POX, GR, and G-S-T) was highly similar in brain, and gut of males, nymphs, and all tissues of females collected from sites A and B. The level of antioxidant response in brain, and gut tissues of males, in brain, and thoracic muscles tissues of female and in gut tissue of nymph from control site with those from sites A and B. Brain, and gut tissue of males from site C created a separate cluster. In brain, thoracic muscles tissues of females, site C created a separate cluster. Antioxidant enzymes activity in brain and thoracic muscles of female and in gut tissue of nymphs from site C had a separate cluster (Fig. 41).

The interaction analysis using GEE showed a significant influence of distance, sex, and tissues on antioxidants response (Table 20). Also, interaction among distance, developmental stage, and tissues had a significant effect on antioxidant response (Table 21).

Chi-square (x^2) test using cross tabulation was performed to examine independency among ROS concentration, and non-enzymatic, and enzymatic response in brain, thoracic muscles, and gut of male, female, and nymph insect samples. The result showed a weak positive equal to 0.42 among ROS, and antioxidant response in thoracic muscles tissue of female insects ($p < 0.05$) (Table 22). Also, x^2 test was performed to examine independency among macromolecules damage, and non-enzymatic, and enzymatic response in tissues of insect samples. The result showed that there was a negative correlation fluctuation from -0.18 to -0.31 among

115

macromolecules damage, and antioxidant response in brain tissue of female, and nymph respectively ($p < 0.05$) (Table 23).

The result showed that significant level of x^2 was lower than 0.05, and there was almost a negative correlation (for APOX, PPO, POX, GR, and G-S-T), and positive correlation for (SOD, and CAT) among pollutants concentration in air, soil, plant, and water and antioxidant enzyme response (Table 24). While, the combined effect of pollutants concentration in air, soil, plant, and water had a positive strong correlation (0.78, 0.96, 0.91 and 0.98) with antioxidant enzymatic response in brain, and thoracic muscles tissue of male, and female insect in SOD, and CAT activity, respectively. The negative correlation occurred among the combined effect of pollutants concentration in air, soil, plant, and water; and brain, thoracic muscle, and gut tissue of nymph, male insect in APOX, POX, and GR activity (-0.88, -0.91, and -0.90, respectively) ($p < 0.05$) (Table 24).

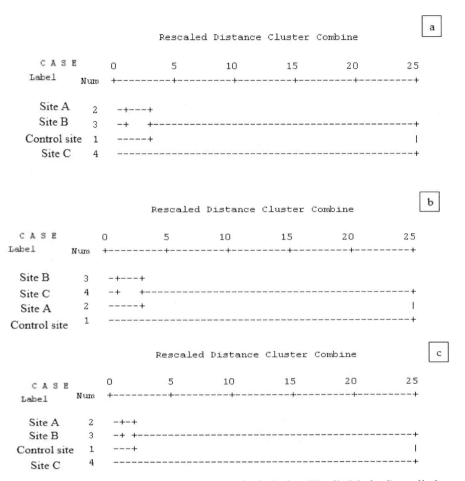

Fig. 39. Dendrogram of the cluster analysis (using Ward's Method) applied for oxidative stress response (non-enzymatic and enzymatic) response (GSH, SOD, CAT, APOX, PPO, POX, GR, G-S-T) in brain (a), thoracic muscles (b), and gut (c) homogenates of male of *Aiolopus thalassinus*, which were collected at control site and polluted sites (A-C) located at different distances from fertilizer factory.

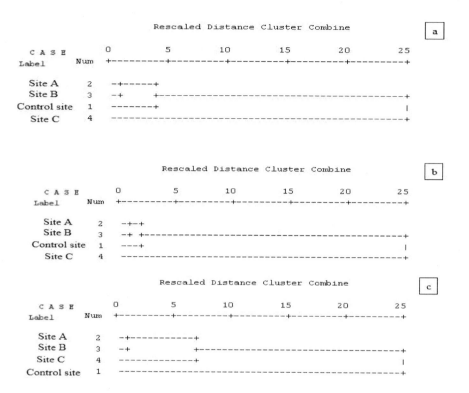

Fig. 40. Dendrogram of the cluster analysis (using Ward's Method) applied for oxidative stress response (non-enzymatic and enzymatic) response (GSH, SOD, CAT, APOX, PPO, POX, GR, G-S-T) in brain (a), thoracic muscles (b), and gut (c) homogenates of female of *Aiolopus thalassinus*, which were collected at control site and polluted sites (A-C) located at different distances from fertilizer factory.

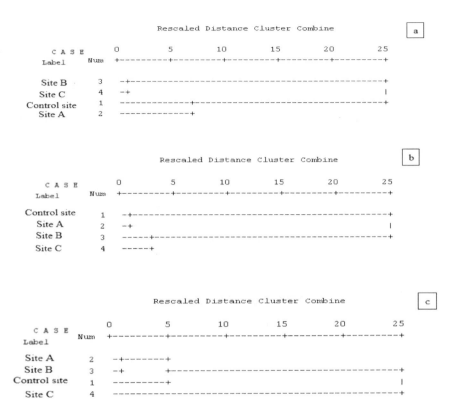

Fig. 41. Dendrogram of the cluster analysis (using Ward's Method) applied for oxidative stress response (non-enzymatic and enzymatic) response (GSH, SOD, CAT, APOX, PPO, POX, GR, G-S-T) in brain (a), thoracic muscles (b), and gut (c) homogenates of 5th instar of *Aiolopus thalassinus*, which were collected at control site and polluted sites (A-C) located at different distances from fertilizer factory.

Table 20. Generalized Estimating Equation to analyze the interactions among the distance from fertilizer factory, types of tissues, sex on oxidative stress response (non-enzymatic and enzymatic response (GSH concentration, and activity of SOD, CAT, APOX, PPO, POX, GR, and G-S-T antioxidant enzymes) in brain, thoracic muscles, and gut homogenates of males, and females of *Aiolopus thalassinus* collected at different sites located at various distances from fertilizer factory.

Source	Chi-square (χ^2)	df	p value
Distance – sex interaction			
GSH	81	2	<0.0001
SOD	694	2	<0.0001
CAT	556	2	<0.0001
APOX	128	2	<0.0001
PPO	52	2	<0.0001
POX	1813	2	<0.0001
GR	4041	2	<0.0001
G-S-T	208	2	<0.0001
Distance – tissue interaction			
GSH	424	4	<0.0001
SOD	625	4	<0.0001
CAT	686	4	<0.0001
APOX	522	4	<0.0001
PPO	474	4	<0.0001
POX	235	4	<0.0001
GR	1302	4	<0.0001
G-S-T	143	4	<0.0001
Sex – tissue interaction			
GSH	169	2	<0.0001
SOD	576	2	<0.0001
CAT	14	2	<0.0001
APOX	131	2	<0.0001
PPO	335	2	<0.0001
POX	1442	2	<0.0001
GR	2509	2	<0.0001
G-S-T	249	2	<0.0001
Distance – tissue – sex interaction			
GSH	371	4	<0.0001
SOD	1244	4	<0.0001
CAT	355	4	<0.0001
APOX	47	4	<0.0001
PPO	204	4	<0.0001
POX	488	4	<0.0001
GR	1484	4	<0.0001
G-S-T	326	4	<0.0001

Table 21. Generalized Estimating Equation to analyze the interactions among the distance from fertilizer factory, types of tissues, developmental stage (adult and nymph) on oxidative stress response (non-enzymatic and enzymatic response (GSH concentration, and activity of SOD, CAT, APOX, PPO, POX, GR, and G-S-T antioxidant enzymes) in brain, thoracic muscles, and gut homogenates of males, females, and 5th instar of *Aiolopus thalassinus* collected at different sites located at various distances from fertilizer factory.

Source	Chi-square (χ^2)	df	p value
Distance – developmental stage interaction			
GSH	75	2	<0.05
SOD	62	2	<0.0001
CAT	99	2	<0.0001
APOX	8	2	<0.0001
PPO	61	2	<0.0001
POX	48	2	<0.0001
GR	416	2	<0.0001
G-S-T	32	2	<0.0001
Distance – tissue interaction			
GSH	27	4	<0.0001
SOD	10	4	<0.05
CAT	26	4	<0.0001
APOX	31	4	<0.0001
PPO	95	4	<0.0001
POX	54	4	<0.0001
GR	132	4	<0.0001
G-S-T	10	4	<0.05
Developmental stage – tissue interaction			
GSH	24	2	<0.0001
SOD	16	2	<0.0001
CAT	49	2	<0.0001
APOX	105	2	<0.0001
PPO	255	2	<0.0001
POX	30	2	<0.0001
GR	92	2	<0.0001
G-S-T	107	2	<0.0001
Distance – tissue – Developmental stage interaction			
GSH	28	4	<0.0001
SOD	16	4	<0.05
CAT	98	4	<0.0001
APOX	323	4	<0.0001
PPO	123	4	<0.0001
POX	36	4	<0.0001
GR	30	4	<0.0001
G-S-T	16	4	<0.05

Table 22. Chi-square (x^2) using cross tabulation was performed to examine independency among reactive oxygen species amount (ROS) (production rate of superoxide anion (O_2^-) and concentration of hydrogen peroxide (H_2O_2)) and oxidative stress response (non-enzymatic and enzymatic) response (GSH, SOD, CAT, APOX, PPO, POX, GR, G-S-T) in brain, thoracic muscles, and gut homogenates of males, females, and 5th instar of *Aiolopus thalassinus*. Samples collected at polluted sites (A-C) located at different distances from fertilizer factory.

	Tissue	X^2	LR	*df*	LLA	*df*	R	r	T	*p* value
Male	Brain	504	147	483	1.93	1	-0.29	-0.46	17.6	< 0.0001
	Thoracic muscle	504	147	483	1.10	1	0.22	0.06	17.6	< 0.0001
	Gut	528	149	506	0.09	1	-0.06	-0.4	19.05	< 0.0001
Female	Brain	456	140	440	1.61	1	-0.26	-0.37	11.4	< 0.0001
	Thoracic muscle	462	141	440	0.91	1	0.20	0.42	9.6	< 0.0001
	Gut	528	149	506	0.32	1	-0.11	0.21	19.05	< 0.0001
Nymph	Brain	438	138	420	1.71	1	0.27	0.27	12.5	< 0.0001
	Thoracic muscle	528	149	506	0.74	1	-0.18	-0.11	19.05	< 0.0001
	Gut	528	149	506	1.71	1	0.24	-0.003	19.05	< 0.0001

X^2, LR, LLA, R, r, and T represent of Person's chi-square, Likelihood Ratio, Linear by Linear Association, Person's R, correlation, and T of Lambda respectively. x^2 revealed was made for each tissue (brain, thoracic muscles, and gut), each sex and developmental stage (male, female, and nymph) separately.

Table 23. Chi-square (x^2) using cross tabulation was performed to examine independency among macromolecules damage (DNA strand breaks, protein carbonyls amount, and lipid peroxides concentration) and oxidative stress response (non-enzymatic and enzymatic) response (GSH, SOD, CAT, APOX, PPO, POX, GR, G-S-T) in brain, thoracic muscles, and gut homogenates of males, females, and 5th instar of *Aiolopus thalassinus*. Samples collected at polluted sites (A-C) located at different distances fertilizer factory.

	Tissue	X^2	LR	*df*	LLA	*df*	R	r	T	*p* value
Male	Brain	4680	599	4623	5.4	1	-0.27	-0.25	37.7	< 0.0001
	Thoracic muscle	4536	593	4488	5.9	1	-0.29	-0.29	32.3	< 0.0001
	Gut	4536	592	4488	5.9	1	-0.28	-0.17	32.1	< 0.0001
Female	Brain	4680	599	4624	2.4	1	-0.18	-0.15	37.7	< 0.0001
	Thoracic muscle	4752	601	4692	1.4	1	-0.14	-0.30	39.1	< 0.0001
	Gut	4680	597	4620	3.3	1	-0.21	-0.35	33.9	< 0.0001
Nymph	Brain	3516	547	3540	7.0	1	-0.31	-0.51	20.3	< 0.0001
	Thoracic muscle	3500	541	3456	4.97	1	-0.26	-0.52	19.1	< 0.0001
	Gut	4266	581	4221	2.89	1	-0.20	-0.41	29.1	< 0.0001

X^2, LR, LLA, R, r, and T represent of Person's chi-square, Likelihood Ratio, Linear by Linear Association, Person's R, correlation, and T of Lambda respectively. x^2 revealed was made for each tissue (brain, thoracic muscles, and gut), each sex and developmental stage (male, female, and nymph) separately.

Table 24. Chi-square (x^2) using cross tabulation was performed to examine independency among air (SO_2, and TSP), soil, plant, and water pollutants, each source separately and combined, concentration (heavy metals (Cu, Zn, Pb, and Cd), and PO_4^{3-}, and SO_4^{2-}) and enzymatic response (SOD, CAT, APOX, PPO, POX, GR, G-S-T) in brain, thoracic muscles, and gut homogenates of males, females, and 5th instar of *Aiolipus thalassinus*. Samples collected at polluted sites (A-C) located at different distances from fertilizer factory.

OR	Pollutant source		Air	Soil	Plant	Water	Pollutants*	*p* value
		Tissue	R	R	R	R	R	
SOD	Male	Brain	0.94	0.87	0.67	0.87	0.78	< 0.05
		Thoracic muscle	0.96	0.59	0.06	0.42	0.55	< 0.05
		Gut	0.95	0.92	0.61	0.85	0.50	< 0.05
	Female	Brain	-0.96	-0.20	0.07	-0.03	0.20	< 0.05
		Thoracic muscle	0.99	0.75	0.45	0.75	0.96	< 0.05
		Gut	0.94	0.42	0.24	0.51	0.90	< 0.05
	Nymph	Brain	-0.98	-0.81	-0.39	-0.72	-0.79	< 0.05
		Thoracic muscle	-0.45	-0.17	-0.21	-0.08	0.54	< 0.05
		Gut	0.96	0.36	0.18	0.35	0.52	< 0.05
CAT	Male	Brain	0.96	0.72	0.32	0.68	0.91	< 0.05
		Thoracic muscle	0.84	0.16	-0.25	-0.04	0.00#	< 0.05
		Gut	0.97	0.92	0.57	0.87	0.77	< 0.05
	Female	Brain	0.98	-0.26	-0.64	-0.39	0.24	< 0.05
		Thoracic muscle	0.98	0.46	0.03	0.39	0.88	< 0.05
		Gut	0.98	0.59	0.24	0.58	0.98	< 0.05
	Nymph	Brain	-0.55	-0.73	-0.74	-0.75	-0.35	< 0.05
		Thoracic muscle	0.01	-0.63	-0.77	-0.74	-0.34	< 0.05
		Gut	0.81	-0.53	-0.77	-0.68	-0.25	< 0.05
APOX	Male	Brain	0.52	-0.28	-0.46	0.44	-0.52	< 0.05
		Thoracic muscle	0.92	-0.25	-0.59	0.43	-0.04	< 0.05
		Gut	-0.83	0.43	0.66	0.47	-0.35	< 0.05
	Female	Brain	-0.92	-0.78	-0.85	-0.84	-0.23	< 0.05
		Thoracic muscle	-0.68	-0.72	-0.71	-0.71	-0.07	< 0.05
		Gut	-0.57	-0.08	0.04	-0.15	-0.66	< 0.05
	Nymph	Brain	-0.94	-0.55	-0.44	-0.63	-0.88	< 0.05
		Thoracic muscle	-0.96	-0.42	-0.03	-0.37	-0.79	< 0.05
		Gut	-0.96	-0.97	-0.78	-0.94	-0.56	< 0.05
PPO	Male	Brain	-0.95	0.39	0.62	0.42	-0.40	< 0.05
		Thoracic muscle	-0.86	0.35	0.64	0.42	-0.35	< 0.05
		Gut	-0.78	0.46	0.60	0.44	-0.38	< 0.05
	Female	Brain	-0.95	0.25	0.54	0.29	-0.52	< 0.05
		Thoracic muscle	-0.94	-0.11	0.31	-0.07	-0.73	< 0.05
		Gut	-0.97	-0.79	-0.46	-0.77	-0.94	< 0.05
	Nymph	Brain	-0.91	-0.72	-0.43	-0.58	-0.22	< 0.05
		Thoracic muscle	-0.94	-0.61	-0.55	-0.51	-0.01	< 0.05
		Gut	-0.95	-0.25	-0.05	-0.08	0.29	< 0.05

R, and MD represent Person's correlation, and enzymatic response (SOD, CAT, APOX, PPO, POX, GR, G-S-T), respectively. * pollutants mean the combined effect of air, soil, plant, and water. ** combined means combined effect of pollutants on DNA, protein, and lipid. # means that $p>0.05$. x^2 revealed was made for each tissue (brain, thoracic muscles, and gut), each sex and developmental stage (male, female, and nymph) separately.

Continue Table 24. Chi-square (x^2) using cross tabulation was performed to examine independency among air (SO_2, and TSP), soil, plant, and water pollutants, each source separately and combined, concentration (heavy metals (Cu, Zn, Pb, and Cd), and PO_4^{3-}, and SO_4^{2-}) and enzymatic response (SOD, CAT, APOX, PPO, POX, GR, G-S-T) in brain, thoracic muscles, and gut homogenates of males, females, and 5th instar of *Aiolopus thalassinus*. Samples collected at polluted sites (A-C) located at different distances from fertilizer factory.

ER		Pollutant source	Air	Soil	Plant	Water	Pollutants*	p value
		Tissue	R	R	R	R	R	
POX	Male	Brain	-0.86	0.19	0.43	0.18	-0.60	< 0.05
		Thoracic muscle	-0.95	-0.32	-0.13	-0.37	-0.91	< 0.05
		Gut	-0.71	0.13	0.28	0.10	-0.65	< 0.05
	Female	Brain	-0.97	0.01	0.25	-0.01	-0.75	< 0.05
		Thoracic muscle	-0.97	0.28	0.46	0.25	-0.55	< 0.05
		Gut	-0.97	-0.19	0.04	-0.23	-0.87	< 0.05
	Nymph	Brain	0.97	-0.36	-0.14	-0.41	-0.89	< 0.05
		Thoracic muscle	0.97	0.02	0.12	-0.76	-0.70	< 0.05
		Gut	-0.98	0.06	0.25	0.02	-0.71	< 0.05
GR	Male	Brain	0.87	0.77	0.85	0.79	0.05	< 0.05
		Thoracic muscle	0.96	0.59	0.64	0.53	-0.22	< 0.05
		Gut	-0.97	-0.42	-0.26	-0.47	-0.90	< 0.05
	Female	Brain	-0.98	0.01	0.44	0.13	-0.50	< 0.05
		Thoracic muscle	-0.98	-0.16	0.29	0.01	-0.38	< 0.05
		Gut	-0.97	-0.64	-0.17	-0.48	-0.57	< 0.05
	Nymph	Brain	-0.58	0.30	0.40	0.24	-0.52	< 0.05
		Thoracic muscle	-0.96	0.18	0.41	0.16	-0.63	< 0.05
		Gut	-0.84	0.23	0.36	0.18	-0.58	< 0.05
G-S-T	Male	Brain	-0.71	0.31	0.45	0.27	-0.51	< 0.05
		Thoracic muscle	-0.93	-0.29	0.11	-0.25	-0.86	< 0.05
		Gut	-0.95	0.29	0.51	0.31	-0.47	< 0.05
	Female	Brain	-0.38	0.58	0.85	0.62	-0.15	< 0.05
		Thoracic muscle	-0.96	-0.41	0.08	-0.25	-0.56	< 0.05
		Gut	-0.97	-0.76	-0.75	-0.83	-0.60	< 0.05
	Nymph	Brain	-0.97	-0.55	-0.12	-0.40	-0.45	< 0.05
		Thoracic muscle	-0.98	-0.76	-0.41	-0.72	-0.87	< 0.05
		Gut	-0.95	-0.64	-0.22	-0.53	-0.87	< 0.05
**Combined	Male	Brain	-0.96	0.002	0.20	0.25	-0.10	< 0.05
		Thoracic muscle	-0.97	0.06	0.57	0.54	-0.18	< 0.05
		Gut	-0.97	-0.06	0.17	0.19	-0.13	< 0.05
	Female	Brain	-0.98	0.02	0.41	0.34	-0.09	< 0.05
		Thoracic muscle	-0.97	-0.06	0.34	0.27	-0.15	< 0.05
		Gut	-0.98	0.08	0.39	0.37	-0.13	< 0.05
	Nymph	Brain	0.94	0.02	0.52	0.42	-0.12	< 0.05
		Thoracic muscle	0.20	0.07	0.52	0.47	-0.13	< 0.05
		Gut	0.96	0.07	0.50	0.45	-0.12	< 0.05

R, and OR represent Person's correlation, and enzymatic response (SOD, CAT, APOX, PPO, POX, GR, G-S-T), respectively. * pollutants mean the combined effect of air, soil, plant, and water. ** combined means combined effect of pollutants on DNA, protein, and lipid. x^2 revealed was made for each tissue (brain, thoracic muscles, and gut), each sex and developmental stage (male, female, and nymph) separately.

125

4 Discussion

In the present work, part of biomonitoring program designed to evaluate the biochemical changes in insect, *Aiolopus thalassinus*, collected from different sites around Abu-Zaabal Company for Fertilizers and Chemical Industries (AZFC). The activity of fertilizer industry has a significant effect on its surroundings, however the air pollutants (SO_2, and PM_{10} for ambient air quality network; and SO_2, and TSP for stack emission of AZFC), heavy metal (Cu, Zn, Pb, and Cd), phosphate (PO_4^{3-}), and sulphate (SO_4^{2-}) concentrations in soil, plant, and water samples from all experimental sites was generally less than the standard limits (US EPA, 2014).

The biomonitoring programs include bioaccumulation, biochemical alterations, morphological and behavior observation, population- and community-level approaches. It's applications are evaluation of metal concentration, toxicology prediction and researches on toxicological mechanism, toxicological prediction and bioremediation (Blackmore and Wang, 2004; Zhou et al., 2008). Toxicological mechanism included detecting the action mechanism through interaction between pollutants and biological macromolecules such as proteins, enzymes and nucleic acid (Markert, et al., 2003). The suitable bioindicators groups which capable of reflecting the pollutants in the ecosystem (Klein, 1999). Proper quality management system (QA) should be applied to make sure that biomonitoring studies are scientifically valid, comparable and robust to evaluate the relationship between environmental pollutants and its toxicological effect in living organism.

Discussion

The results of time series modelling of average concentrations of ambient air in AZFC (Fig. 2 and 5) showed that the time series had not have a seasonal fluctuation with somewhat a constant trend which related to the behaviour of PACF, ACF and their residuals (Gong et al., 2013). These results lead to a suggestion that the temporal analysis of these data to monitor environmental pollution in AZFC is not feasible. So, the using of temporal analysis of pollutants concentration to biomonitor environmental pollution was performed in our study (Naderizadeh et al., 2016).

The result of person's R (in chi-square test) between SO_2, and PM_{10} (for ambient air) was -0.12, SO_2, and TSP (for stack emission) was -0.22 ($p < 0.05$). This significantly negative correlation was previously reported by Mansouri, et al. (2011), and explained by Favez et al. (2008) who referred the reason of negative correlation between these pollutants due to sinking properties of dust particles for acidic gaseous species, such as SO_2, HNO_3 and HCl. This means that, dust particles can adsorb acidic gaseous species on its surface. Otherwise, chemical composition of particles depends on geographical, meteorological, and source-specific variables. Generally, it consists of biological components, organic and inorganic compounds (Lodovici, and Bigagli, 2011; Fuzzi, et al., 2015).

In the present work, ambient air pollutants, stack emission pollutants (hydrofluoric acid fumes HF, SO_2, NO_2, PM_{10}, waste products of fertilizer industry such as phosphates, sulphates, dust, and heavy metals) increased the production of ROS in cells of exposed organisms and caused oxidative stress. The pollutants which used in industry and agriculture include metals, metalloids, and numerous

127

other organic compounds, lead to increase the production of reactive oxygen species (ROS) in the cells of individuals exposed to them, and therefore caused oxidative stress with all the adverse consequences for organisms (Farahat et al., 2010; Okamoto et al., 2014; Zhu et al., 2014; Shinkai et al., 2015; Yousef et al., 2017; Abdelfattah et al., 2017). This might explain the fluctuating results among the different sites. Ahmad (1995) and Chaitanya et al. (2016) approved that, the pollution of environmental components (air, soil, water and plant) were considered as a non-biological factor of oxidative stress in invertebrates.

Control and polluted sites differed among each other in the soil, plant, and water content of heavy metals (Pb, Cd, Zn, and Cd), but above all, they significantly varied in PO_4^{3-} as well as SO_4^{2-} concentration in the soil, plant, and water samples (Fig. 8-10). Phosphorus species were considered the principal carriers of trace elements in soils. The phosphate industry poseses a serious soil pollution hazard, with deposited contaminants being potentially hazardous to plants and groundwater (Kassir et al., 2012), and this consequently affected the plant, and the living organism inhibited in this soil, and water (Table 6 and 7).

The concentrations of heavy metals, sulphates and phosphates were almost usually higher in soil, plant, and water samples collected from polluted sites comparing to the control site (Fig 8-10). At site A the concentrations of Zn, Cu, Cd, Pb and sulphates achieved the highest values among all sites (Table 2). However, the relationship between the level of contaminants in soil, plant, and water samples and the distances of the sites from source of pollution was not clear. Some differences may result from composition, even subtle, of the soil at all sites. Components such as clay, minerals and organic substances may change the dynamic of pollutants uptake (Wuana and Okieimen, 2011). The exception of high concentration of PO_4^{3-} in plant samples than soil, and water samples was explained by Gheorghe and Ion (2011). They mentioned that industrial dust blocks stomata and lowers their conductance to CO_2, and inhibit photosystem, when acidic air gases enter leaves through stomata by diffusion pathway, these acidic gases dissolves in cells and therefore, gives rise to phosphate, sulphate, and nitrite ions which are toxic at high concentrations.

The strong positive correlation between air pollutants and SO_4^{2-} in soil, plant, and water (Table 5) supported the results of Hand et al. (2012) who reported that annual mean sulfate concentration at UK rural sites were declined at 2.7% per year over the period 1992–2010, with a linear relationship between SO_2 emissions and sulfate concentrations.

In the present work, the level of stress was evaluated indirectly, by assessing reactive oxygen species concentration ($O_2^{\cdot-}$ and H_2O_2) (Fig 13 and 14), oxidative damage of macromolecules: DNA (Fig. 18-22), proteins (Fig. 26) and lipids (Fig. 27), as well as measurement of non-enzymatic (GSH concentration) (Fig. 31), and enzymatic antioxidant (SOD, CAT, APOX, PPO, POX, GR, and G-S-T) (Fig. 32-38) response in brain, thoracic muscles and gut of *A. thalassinus*. This was done in accordance with generally accepted knowledge (Fang et al., 2002; Halliwell and Whiteman, 2004). Heavy metals tend to accumulate in an organism. They can indirectly increase the production of ROS (such as $O_2^{\cdot-}$, and H_2O_2) in the cells by affecting the rate of respiratory metabolism for which oxygen is the most essential factor (Azam et al., 2015).

Generally, the level of DNA damage in tissues of *A. thalassinus* from the polluted sites was significantly higher than in individuals from control sites (Fig 18-22). However, no single parameter of the comet did reflect the level of contamination to the same extent as a comprehensive analysis of all DNA damage parameters. The obtained results indicated a possible impact of the pollutants of fertilizer company on DNA integrity in *A. thalassinus* tissues. Therefore, the potential use of the comet assay as a biomonitoring method of the environmental pollution, caused by fertilizer industry, should be considered (Lovell and Omori, 2008; Valverde and Rojas, 2009; Augustyniak et al., 2016a).

The present results are in a compliance with the data presented by Lobo et al. (2010), who reported that DNA is one of the most important targets of free radical attack in the cells. Molecular mechanisms causing DNA damage may include: activation of nucleases, direct reaction of free radicals such as hydroxyl radicals ($^{\bullet}$OH) with the DNA, or breaks resulting from free radical reaction of deoxyribose residues, invariably possess blocked termini (Halliwell and Aruoma, 1991; Hegde et al., 2008). Moreover, DNA damage can involve: formation of a basic site that finally leads to strand breaks (Halliwell, 1999; Friedberg et al., 2005; Cakmakoglu et al., 2011); modifications and degradations of nitrogenous bases; damage to sugar moiety; formation of DNA-DNA and DNA-protein cross links, and damage to the repairing system of DNA (Kohen and Nyska, 2002; Birben et al., 2012).

Discussion

It was proved that metals such as Zn caused DNA damage in cells isolated from the brain of *Chorthippus brunneus*, but this effect was not proportional to the metal dose (Augustyniak et al., 2006). Copper, another essential trace element, can induce oxidative damage to macromolecules such as DNA, proteins and lipids (Shukla et al., 2011). In the presence of the reducing agents, Cu can catalyze the production of ROS, such as superoxide anion radical ($O_2^{\cdot-}$), hydrogen peroxide (H_2O_2) and $^{\cdot}OH$, through Fenton and Haber-Weiss reactions. Previous studies showed that DNA breaks are caused by a site-specific reaction of Cu ions, both *in vitro* and *in vivo* (Hayashi et al., 2000).

Yousef et al. (2010) found that the genotoxicity of heavy metals, cadmium and lead, in *Schistocerca gregaria* was very high. Hence, this may reflect the role played by *S. gregaria* as a valuable bioindicator of environmental genotoxic pollutants. Joseph (2009) hypothesized that genotoxic effect of Cd may also result from generation of ROS, and lead to oxidative stress that is associated with generation of 7,8-dihydro-8-oxoguanine (8-oxoGua) commonly used to monitor DNA damage (Shukla et al., 2011).

Some studies proved that inhalation of SO_2 caused a significant increase in DNA damage in various organs of both males and females of albino mice which could cause mutation, cancer, and other diseases related to the DNA damage (Meng et al., 2005). Severe intensification of oxidative stress and the increase in DNA damage – occurring due to PM exposure – were described (Prahalad et al., 2001; Rhoden et al., 2005; Gurjar et al., 2010).

The relationship between comet parameters and distance of sites from AZFC in the present research has no clearly definite pattern

(Table 9), however, the observed strong negative correlation between % severed cells and the distance of the polluted sites from AZFC. Based on the highly significant correlation confirmed in brain, thoracic muscles, and gut of males and females collected from polluted and control sites (Table 9) and the high level of similarity among insects from all the polluted sites (Fig. 23-25), it was suggested that specific pollution resulting from the activity of the fertilizer industry can cause comparable adverse effects in the organisms inhabiting areas up to 6 km from the source of contamination. It is concluded that the percentage of the cells with visible DNA damage (% of severed cells) was the best parameter for monitoring of fertilizer pollutants. It was probably correlated with an extremely high concentration of phosphates and sulphates in all the polluted sites.

The present result revealed the spectacular elevation of protein carbonyls and lipid peroxides in tissues of insects collected at polluted sites compared to the control insects. It was observed in insects from all contaminated sites (Fig. 26 and 27), strong negative correlations between lipid peroxides in all tissues of male, female, and nymph and the distance of the insect collection sites away from AZFC were stated (Table 12). The previous results showed the level of lipid peroxides may serve as a good marker of oxidative stress in areas contaminated by fertilizer industry (Lushchak, 2011).

The present result showed that the concentration of protein carbonyls amount and lipid peroxide concentration were higher in the of 5th instar tissues than adult (male and females) tissues, in samples collected from site A (Fig. 26 and 27). Some insects can eliminate heavy metals from their bodies through ecdysis and metamorphosis

(Maroni and Watson, 1985), and it was some studies found that heavy metal concentration in the feces increased with grasshopper development (Zhang et al., 2011).

The obtained results proved the information that proteins are important targets of free radical attack in the cells (Lushchak, 2011), and thus the antioxidant defense, cellular function, and finally organism survival can be impaired. ROS are known to convert amino groups of proteins and thereby, change protein structure and function. The oxidative stress increased the number of modified carbonyl groups correlates with protein damage (Hermes-Lima, 2004). Also, ROS can cause fragmentation of the peptide chain, alteration of electrical charge of proteins, cross-linking of proteins, and oxidation of specific amino acids and therefore lead to increased susceptibility to proteolysis by degradation of specific proteases (Kelly and Mudway, 2003). The oxidation of proteins leads physiologically to disruption of conformation and vital functions of protein molecules, including enzymes, and other regulatory functions of the cell (Korsloot et al., 2004; Birben et al., 2012).

Lipid peroxidation usually measured as a level of lipid peroxides has been used frequently to analyze the effect of pollutants (Livingstone, 2001; Lushchak, 2007; 2011). Lipid peroxidation products, such as isoprostanes and thiobarbituric acid reactive substances, are used as indirect biomarkers of oxidative stress (Birben et al., 2012). As in the case of protein carbonyls, the highest concentration of lipid peroxides was observed in tissues of individuals from site A, the nearest to the source of contamination. Significant oxidative damage, including lipid peroxidation occurred if antioxidant

133

defense systems are overwhelmed by ROS production (Winston, 1991; Halliwell and Gutteridge, 2015).

Increased concentration of ROS usually increases the activity of antioxidant enzymes, like superoxide dismutase (SOD), catalase (CAT), and peroxidases (Table 22). There are key antioxidant enzymes responsible for scavenging of oxygen radicals (Donahue et al., 1997; Khaper et al., 2003; Halliwell and Gutteridge, 2015; Dutta et al., 2016). However, environmental pollutants such as heavy metals increase the production of reactive oxygen species (ROS), and, directly or indirectly, cause oxidative damage by inhibiting activity of antioxidant enzymes. Previous studies suggested that inhibition of antioxidant enzymes activity occurred due to decrease expression level of antioxidant enzymes (Chaitanya et al., 2014). In *Oxya chinensis*, high concentrations of Cd acted directly, causing increase in reactive oxygen species, and changes in SOD, CAT, APOX, PPO, POX G-S-T, GR activity (Lijun et al. 2005; Shukla et al., 2017). The authors suggested that multiple mechanisms rather than a single mechanism may be responsible for the capacity of insects to resist cadmium. Moreover, they concluded that CAT has a strong detoxification function and play the most important role in limitation of the damaging effects of reactive oxygen species in *O. chinensis* injected with cadmium (Lijun et al. 2005). Positive correlation between CAT activity and distance of sites from the fertilizer company was described. A highly significant correlation was confirmed in male brains as well as female and male gut (Table 18). Moreover, the activity of CAT was significantly lower in brain, thoracic muscles, and gut of both male and female *A. thalassinus* collected from polluted sites, comparing to control individuals (Fig. 33).

Discussion

Catalase scavenges H_2O_2 at high concentrations, whereas ascorbate peroxidase scavenges H_2O_2 at low concentrations, the negative significant correlation between enzymes activity was revealed (Table 19). The key role of CAT in ROS scavenging, also in plants, was studied (Sofo et al, 2015). APOX, and POX catalyzes the reduction of H_2O_2 with consumption of ascorbate as the reducing agent. Therefore, APOX activity depends exclusively on the availability of reduced ascorbate. Under normal conditions the cellular pool of ascorbate is kept in a reduced state by a set of enzymes, namely mono-dehydroascrobate reductase (MDAR) and dehydroascorbate reductase (DHAR) capable of using NAD(P)H to regenerate oxidized ascorbate (Farooqui and Farooqui, 2011). The present work confirmed the positive correlation between APOX activity in gut of both sexes, and the distance of insect collection sites from the pollution source. It revealed that the oxidative stress markers of gut homogenates in comparison to other tissues were frequently higher in insects collected from different experimental sites. This is probably because the gut of this insect is usually subjected to prooxidants ingested in food, as reported in other phytophagous insects (Ahmad, 1992; Felton and Summers, 1995; Krishnan and Kodrik, 2006).

The present result, usually showed that the activity of antioxidant enzymes (SOD, CAT, APOX, PPO, POX G-S-T, GR) in 5th instar tissues were lower than adult (male and female) tissues (Fig. 32-38). It was suggested that the depletion of the antioxidant enzyme activity in 5th instar may be due to accumulation of heavy metals in nymphs more than adult, which lead to inactivate enzyme activity as result of structural changes of the antioxidant enzyme (Iszard ,1995; Wilczek et al. 2003; Yan et al., 2007). Also, Zhang et al. (2011)

135

referred the depletion of antioxidant enzyme activity occurred due to decrease in protein synthesis which depend on aging (developmental stage). Woodring and Sparks, (1987) suggested that the developmental stage-dependent changes in antioxidant enzyme activity may occur as result of differences in the expression levels of the isoenzymes which are involved in the endogenous substrates transformation during development.

The results in Table 19, showed a significant relationship between SOD and CAT antioxidant enzymes in brain, thoracic muscles, and gut of male, female, and 5th instar nymphs of $A.$ $thalassinus$. Felton and Summers, (1995) explained this relationship as SOD can convert the free superoxide radical ($O_2^{\bullet-}$) to H_2O_2, which is then eliminated by CAT.

The comparison between CAT and APOX in the same tissue (Fig. 33 and 34) leads to the suggestion that APOX has an important role in the analyzed grasshopper populations, under constant environmental stress. Therefore, the previously mentioned phenomenon of CAT which act lengthy at higher concentrations of H_2O_2 than APOX, and POX lead to a suggestion of presence other efficient mechanisms for countering the effects of xenobiotics before a strong oxidative stress occurs, observable as for example a sudden increase of H_2O_2.

Research conducted in this study suggests that significant differences of environmental stress marker levels in $A.$ $thalassimus$ are not a direct result of the site of insect collection, but depend on differences in contamination among studied sites. Similar results and finding were observed by Ihechiluru et al., (2015), who found insignificant difference between oxidative stress markers and the sites of collections, however there were strong positive or negative overall correlations between heavy metal concentrations in insects and respective oxidative stress markers.

5 Conclusion

Phosphate fertilizer industries produce environmental pollutants such as hydrofluoric acid (HF), sulphur dioxide (SO_2), nitrogen dioxide (NO_2), particulate matter of 10 mm diameter (PM_{10}), and dust. Transport and deposition of such pollutants may have a hazardous effect on the environment, particularly air, soil, plant, and water. The results revealed that, phosphate fertilizer industry enhances increase of pollutants concentration in plant than in soil, and water.

These environmental pollutants primarily increased the production of reactive oxygen species (ROS) such as superoxide anion ($O_2^{\cdot-}$), hydroxyl radical ($^{\cdot}OH$) and hydrogen peroxide (H_2O_2). The ROS lead to oxidative stress, and therefore, damaging the macromolecules (DNA single strand breaks, protein carbonyls, and lipid peroxides) when exceed normal level.

Biomonitoring program includes planning phase (which include legislative rules, goals, objectives, selection target population, identification of stakeholders, and ethical consideration) and implementation phase (which include data management, bio specimen, collection, lab analysis, statistical analysis, and results interpretation). Our study focused on biochemical alterations which has toxicological prediction applications. Toxicological mechanism included detecting the action mechanism through interaction between pollutants and biological macromolecules such as proteins, enzymes and nucleic acid. The suitable bioindicators groups which capable of reflecting the pollutants present in the ecosystem, for example, *Aiolopus thalassinus* which has terrestrial habitat, and high sensitivity to environmental pollutants. Our results showed that, adult female was more protective against environmental stress than adult male and 5th instar insect. The

138

susceptibility to environmental stress in insect tissues increased in the order of brain, thoracic muscles, and gut.

Protection against environmental stress can be realized by two main mechanisms: (i) the avoidance of stress, which cannot be achieved by organisms living in polluted areas or (ii) the intensification of the antioxidative defense of the organism. Antioxidants response includes non-enzymatic (such as glutathione reduced (GSH)), and enzymatic antioxidants (such as superoxide dismutase (SOD), catalase (CAT), glutathione peroxidase (GPx), ascorbate peroxidase (APOX), polyphenoloxidase (PPO), glutathione reductase (GR), and glutathione-s-transferase (GST)) are perceived to be important indicators of oxidative stress.

7 References

Abdelfattah, E. A., Augustyniak, M., and Yousef, H. A. (2017). Biomonitoring of genotoxicity of industrial fertilizer pollutants in *Aiolopus thalassinus* (Orthoptera: Acrididae) using alkaline comet assay. *Chemo.*, *182*, 762-770.

Aebi, H. (1984). Catalase *in vitro*. *Method Enzymol. 105*: 121-126.

Ahmad, S. (1992). Biochemical defense of pro-oxidant plant allelochemicals by herbivorous insects. *Biochem. Syst. Ecol. 20(4):* 269-296.

Ahmad, S. (1995). Oxidative stress from environmental pollutants. *Archi. of insect biochem. and physio., 29(2),* 135-157.

Allen, R. G., Farmer, K. J., Newton, R. K., and Sohal, R. S. (1984). Effects of paraquat administration on longevity, oxygen consumption, lipid peroxidation, superoxide dismutase, catalase, glutathione reductase, inorganic peroxides and glutathione in the adult housefly. *Compar. Biochem. and Physi. Part C: Comparative Pharmacology*, *78*(2), 283-288.

Al-Shami, S. A., Rawi, C. S., Ahmad, A. H., Nor, S. A., (2012). Genotoxicity of heavy metals to the larvae of *Chironomus kiiensis Tokunaga* after short-term exposure. *Toxicol. Ind. Health 28 (8),* 734e739. DOI. 0748233711422729.

Amado, L. L., Robaldo, R. B., Geracitano, L., Monserrat, J. M, Bianchini, A. (2006). Biomarkers of exposure and effect in the Brazilian flounder *Paralichthys orbignyanus* (Teleostei: Paralichthyidae) from the Patos Lagoon estuary (Southern Brazil). *Marine pollut Bull. 52(2):* 207-213.

Asada, K. (1984). Chloroplasts: Formation of active oxygen and its scavenging. *Method Enzymol. 105*: 422-429.

Augustyniak, M., Gladysz, M., Dziewie, cka, M., (2016a). The Comet assay in insects -Status, prospects and benefits for science. Mut. Res. Rev. Mut. Res. 767, 67e76.

Augustyniak, M., Juchimiuk, J., Przybyłowicz, W. J., Mesjasz-Przybyłowicz, J., Babczyńska, A., and Migula, P. (2006). Zinc-induced DNA damage and the distribution of metals in the brain of grasshoppers by the comet assay and micro-PIXE. *Comp. Biochem. and Physiol. C: Toxicol. and Pharma, 144(3)*: 242-251.

Augustyniak, M., and Migula, P. (2000). Body burden with metals and detoxifying abilities of the grasshopper—*Chorthippus brunneus* (Thunberg) from industrially polluted areas. In: Merkert, B., Friese, K. (Eds.), Trace Elements—Their Distribution and Effects in the Environment. *Elsevier Sci., Amsterdam*, pp. 423–454

Augustyniak, M., and Migula, P., (1996). Patterns of glutathione S-transferase activity as a biomarker of exposure to industrial pollution in the grasshopper *Chorthippus brunneus* (Thunberg). SSTOR., 4, 9–15.

Augustyniak, M., Płachetka-Bo_zek, A., Kafel, A., Babczy_nska, A., Tarnawska, M., Janiak, A., Loba, A., Dziewie͵cka, M., Karpeta-Kaczmarek, J., Zawisza-Raszka, A., (2016b). Phenotypic plasticity, epigenetic or genetic modifications in relation to the duration of Cd-Exposure within a microevolution time range in the beet armyworm. *PLoS One* 11 (12): e0167371. http://dx.doi.org/10.1371/ journal. pone.0167371.

Azam, I., Afsheen, S., Zia, A., Javed, M., Saeed, R., Sarwar, M. K., and Munir, B. (2015). Evaluating Insects as Bioindicators of Heavy Metal Contamination and Accumulation near Industrial Area of Gujrat, Pakistan. *BioMed. Res. Int., 1-*11. doi: 10.1155/2015/942751

Bilbao, C., Ferreiro, J. A., Comendador, M. A., and Sierra, L. M. (2002). Influence of mus201 and mus308 mutations of *Drosophila melanogaster* on the genotoxicity of model chemicals in somatic cells in vivo measured with the comet assay. *Mut. Res. 503(1):* 11-19.

Birben, E., Sahiner, U. M., Sackesen, C., Erzurum, S., and Kalayci, O. (2012). Oxidative stress and antioxidant defense. The *World Allergy Organ J., 5(1)*: 9-19.

Blackmore, G., and Wang, W. X. (2004). Relationships between metallothioneins and metal accumulation in the whelk *Thais clavigera. Marine Eco. Prog. Series*, *277*, 135-145.

Bonham, C.D. (1989) "Measurement for terrestrial vegetation," John *Wiley and Sons*, New York, NY.

Boon, D. Y., Soltanpour, P. N. (1991). Estimating Total Lead Cadmium and Zinc in Contaminated Soils from Ammonium Bicarbonate 3-Dtpa-Extractable Levels. *Commu Soil Sci Plant Anal. 22*(5-6): 369.

Bradford, M. M. (1976). A rapid and sensitive method for the quantitation of microgram quantities of protein utilizing the principle of protein-dye binding. *Anal Biochem. 72(1):* 248-254.

Cakmakoglu, B., Cincin, Z. B., and Aydin, M. (2011). Effect of Oxidative Stress on DNA Repairing Genes, In: Selected Topics in DNA Repair. *Chen, C.* (Ed.), InTech.

Carlberg, I. and Mannervik, B. (1985). Glutathione reductase assay. *Methods in Enzym. 113:* 484–495.

Carmona, E. R., Creus, A., Marcos, R. (2011). Genotoxicity testing of two lead compounds in somatic cells of *Drosophila melanogaster. Mutat. Res. 724*: 35e40.

Chaitanya, R. K., Shashank, K., and Sridevi, P. (2016). Oxidative Stress in Invertebrate Systems. In *Free Radi. and Disea.* InTech.

Chaitanya, R., P., Sridevi, K. S., Kumar, B. S., Mastan, K. A., and Dutta-Gupta., A. (2014). Expression analysis of reactive oxygen species detoxifying enzyme genes in *Anopheles stephensi* during Plasmodium berghei midgut invasion. Asian *Pacific j. of trop. Med. 7*: 680-684.

Chen, T. B., Zheng, Y. M., Lei, M., Huang, Z. C., Wu, H. T., Chen, H., and Tian, Q. Z. (2005). Assessment of heavy metal pollution in surface soils of urban parks in Beijing, China. *Chemo, 60(4)*: 542-551.

Chen, W. P., and Li, P. H. (2001). Chilling-induced Ca^{2+} overload enhances production of active oxygen species in maize (*Zea mays* L.) cultured cells: the effect of abscisic acid treatment. *Plant, Cell & Enviro. 24*(8), 791-800.

Collins, A., Koppen, G., Valdiglesias, V., Dusinska, M., Kruszewski, M., and Møller, P. (2014). The comet assay as a tool for human biomonitoring studies: The Comet project. *Mutat. Res/ Rev. Mutat. Res. 759*: 27-39. Comet Score Tutorial. ©2013 TriTek Corp. http://AutoComet.com

Dalle-Donne, I., Rossi, R., Giustarini, D., Milzani, A., and Colombo, R. (2003) Protein carbonyl groups as biomarkers of oxidative stress. *Clin Chim Acta 329*:23–38

Dhawan, A., Bajpayee, M., and Parmar, D. (2009). Comet assay: a reliable tool for the assessment of DNA damage in different models. *Cell Biol. Toxicol, 25(1):* 5-32.

Donahue, J. L., Okpodu, C. M., Cramer, C. L., Grabau, E. A., Alscher, R. G. (1997). Responses of antioxidants to paraquat in pea leaves (relationships to resistance). *Plant physiol. 113(1):* 249-257.

Dos -Anjos, N. A., Schulze, T., Brack, W., Val, A. L., Schirmer, K., and Scholz, S. (2011). Identification and evaluation of cyp1a transcript expression in fish as molecular biomarker for petroleum contamination in tropical fresh water ecosystems. *Aquat toxicol.103(1):* 46-52.

Dutta, P., Dey, T., Manna, P., and Kalita, J. (2016). Antioxidant Potential of *Vespa affinis* L., a Traditional Edible Insect Species of North East India. *PloS one. 11(5):* e0156107.

EEA, 2015, The European environment — state and outlook 2015: synthesis report, European Environment Agency, Copenhagen.

Egyptian Industrial Development Authority (IDA). (2014)

Egyptian Pollution Abatement Project (PMU). (2012). http://www.eeaa.gov.eg/portals/0/eeaaReports

ESE (2014). Egypt State of Environment report. Egyptian Environmental Affairs Agency.

Fang, Y.Z., Yang, S., and Wu, G. (2002). Free radicals, antioxidants, and nutrition. *Nutri.18:* 872-879.

Farahat, A. A., Al-Sayed, A. A., and Mahfoud, N. A. (2010). Compost and other organic and inorganic fertilizers in the scope of the root-knot nematode reproduction and control. *Egyptian J Agronematol. 9:* 18-29.

Farooqui, T., and Farooqui, A. A. (2011). Oxidative stress in vertebrates and invertebrates: Molecular aspects of cell signaling. *John Wiley & Sons*.

Favez, O., Cachier, H., Sciare, J., Alfaro, S. C., El-Araby, T. M., Harhash, M. A., and Abdelwahab, M. M. (2008). Seasonality of major aerosol species and their transformations in Cairo megacity. *Atmos. Enviro.*, *42*(7), 1503-1516.

Felton, G. W., and Summers, C. B. (1995). Antioxidant systems in insects. *Arch of insect biochem. and physio.*, *29*(2), 187-197.

Friedberg, E. C., Walker, G. C., Siede W., Wood, R.D., Schultz, R.A., and Ellenberger, T. (2005). DNA Repair and Mutagenesis, 2nd ed. Washington, DC, *USA: ASM Press*.

Fuzzi, S., Baltensperger, U., Carslaw, K., Decesari, S., Denier Van Der Gon, H., Facchini, M. C., and Nemitz, E. (2015). Particulate matter, air quality and climate: lessons learned and future needs. *Atmo. Chem. and physics*, *15(14)*, 8217-8299.

Gheorghe, I. F., and Ion, B. (2011). The effects of air pollutants on vegetation and the role of vegetation in reducing atmospheric pollution. In *The impact of air pollution on health, economy, environment and agricultural sources*. InTech.

Gilbert D L (2000) Fifty years of radical ideas. Ann. *New York Acad Sci. 899(1)*: 1-14.

Gong, S. H., Gao, Y. F., Shi, H. B., and Zhao, G. (2013). A practical MGA-ARIMA model for forecasting real-time dynamic. *Radio Sci., 48, (3), 208-225*.

Guanggang X., Diqiu L., Jianzhong Y., Jingmin G., Huifeng Z., and Mingan S., et al. (2013). Carbamate insecticide methomyl confers cytotoxicity through DNA damage induction. *Food Chem. Toxicol. (53)*, 352–358 10.

Gurjar, B. R., Molina, L. T., and Ojha, C. S. P. (2010). Air pollution: health and environmental impacts. *CRC Press*.

Gutteridge, J. M. (1995). Lipid peroxidation and antioxidants as biomarkers of tissue damage. *Clin Chem 41:*1819–1828.

Gyori, B. M., Venkatachalam G., Thiagarajan P. S., Hsu D., and Clement M. (2014). OpenComet: An automated tool for comet assay image analysis. *Red. Bio.. 2*, 457-465. doi.org/10.1016/j.redox.2013.12.020.

Halliwell, B., and Gutteridge, J. M. (1984) Oxygen toxicity, oxygen radicals, transition metals and disease. *Biochem J. 219(1):*1-14.

150

Halliwell, B., and Gutteridge, J. M. (2015). Free Radical Bio Med. *Oxford University Press, USA.*

Halliwell, B., and Whiteman, M. (2004). Measuring reactive species and oxidative damage *in vivo* and in cell culture: how should you do it and what do the results mean? *British J Pharmacol. 142(2)*: 231-255.

Halliwell, B. (1999). Oxygen and nitrogen are pro-carcinogens. Damage to DNA by reactive oxygen, chlorine and nitrogen species: measurement, mechanism and the effects of nutrition. *Mut Res, 443(1),* 37-52.

Halliwell, B., and Aruoma, O. I. (1991). DNA damage by oxygen-derived species Its mechanism and measurement in mammalian systems. *FEBS letters, 281(1-2)*, 9-19.

Hand, J. L., Schichtel, B. A., Malm, W. C., and Pitchford, M. L. (2012). Particulate sulfate ion concentration and SO2 emission trends in the United States from the early 1990s through 2010 Atmos. *Chem. Phys. 12*: 10353–65

Hayashi, M., Kuge, T., Endoh, D., Nakayama, K., Arikawa, J., Takazawa, A., and Okui, T. (2000). Hepatic copper accumulation induces DNA strand breaks in the liver cells of Long-Evans Cinnamon strain rats. *Biochem. Biophys. Res. Commun. 276(1)*: 174-178.

Hegde, M. L., Hazra, T. K., and Mitra, S. (2008). Early steps in the DNA base excision/single-strand interruption repair pathway in mammalian cells. *Cell Res. 18(1)*: 27-47.

Hermes-Lima, M. (2004). Oxygen in biology and biochemistry: role of free radicals. *Funct. Meta.: Regulation and adaptation. 1:* 319-966.

Hermes-Lima, M., Willmore, W. G., and Storey, K. B. (1995). Quantification of lipid peroxidation in tissue extracts based on Fe (III) xylenol orange complex formation. *Free Rad. Bio Med. 19(3):* 271-280.

Ihechiluru, N. B., Henry, A. N., and Taiwo, I. E. (2015). Heavy metal bioaccumulation and oxidative stress in *Austroaeschna inermis* (Dragon fly) of the Lagos Urban ecosystem. *J Environ Chem. Ecoto.7(1):* 11-19.

Iszard, M. B., Liu, J., and Klaassen, C. D. (1995). Effect of several metallothionein inducers on oxidative stress defense mechanisms in rats. *Toxico., 104*(1-3), 25-33.

Jha, A. N. (2008). Ecotoxicological applications and significance of the comet assay. *Muta. 23(3)*: 207-221.

Joseph, P. (2009). Mechanisms of cadmium carcinogenesis. *Toxicol. Appl. Pharmacol. 238(3):* 272-279.

Junglee, S., Urban, L., Sallanon, H. and Lopez-Lauri, F. (2014) Optimized Assay for Hydrogen Peroxide Determination in Plant Tissue Using Potassium Iodide. *Ameri. J.of Analy. Chem., 5,* 730-736.

Kassir, L. N., Lartiges, B., and Ouaini, N. (2012). Effects of fertilizer industry emissions on local soil contamination: a case study of a phosphate plant on the east Mediterranean coast. *Environ Technol. 33(8):* 873-885.

Kaviraj, A., Unlu, E., Gupta, A., and El Nemr, A. (2014). Biomarkers of environmental pollutants. *Bio. Med. Res. Int.* 2 pages. doi.org/10.1155/2014/806598

Kelly, F. J., Mudway, I. S. (2003). Protein oxidation at the air-lung interface. *Amino Acids. 25(3-4):* 375-396.

Khaper, N., Kaur, K., Li, T., Farahmand, F., Singal, P. K. (2003). Antioxidant enzyme gene expression in congestive heart failure following myocardial infarction. In Biochemistry of Hypertrophy and Heart Failure (pp. 9-15). *Springer US*

Klein, R., (1999). Retrospektive Wirkungsforschung mit lagerfähigen Umweltproben. In: Oehlmann, J., Markert, B. (Eds), Ökotoxikologie – Ökosystemare Ansätze und Methoden, Ecomed, Landsberg, pp. 285–293.

Kohen, R. and Nyska, A. (2002). Oxidation of biological systems: oxidative stress phenomena, antioxidants, redox reactions, and methods of their quantification, *Toxic Path, 30 (6):* 620-50.

Korsloot, A., Van Gestel, C. A., and Van Straalen, N. M. (2004). Environmental stress and cellular response in arthropods. *CRC Press.*

Krishnan, N., and Kodrík, D. (2006). Antioxidant enzymes in *Spodoptera littoralis* (Boisduval): are they enhanced to protect gut tissues during oxidative stress? *J Insect Physiol. 52(1):* 11-20.

Kumar, K. B., and Khan, P. A. (1982). Peroxidase and polyphenol oxidase in excised ragi (Eleusine corocana cv PR 202) leaves during senescence. *Int J exp Biol. 20(5):* 412-416.

Levine, R. L., Garland, D., Oliver, C. N., Amici, A., Climent, I., Lenz, A. G., Stadtman, E. R. (1990). Determination of carbonyl content in oxidatively modified proteins. *Method Enzymol. 186:* 464-78.

Lijun, L., Xuemei, L., Yaping, G., Enbo, M. (2005). Activity of the enzymes of the antioxidative system in cadmium-treated *Oxya chinensis* (Orthoptera Acridoidae). *Environ Toxicol Pharmacol. 20(3):* 412-416.

Livingstone, D. R. (2001). Contaminant-stimulated reactive oxygen species production and oxidative damage in aquatic organisms. *Mar Pollut Bull. 42(8):* 656-666.

Lobo, V., Patil, A., Phatak, A., and Chandra, N. (2010). Free radicals, antioxidants and functional foods: Impact on human health. *Pharma. Revi.*, *4*(8), 118–126.

Lodovici, M., and Bigagli, E. (2011). Oxidative stress and air pollution exposure. *J.of toxic.*,

Lovell, D. P., and Omori, T. (2008). Statistical issues in the use of the comet assay. *Mutage.* 23(3): 171-182.

Lushchak, V. I. (2007). Free radical oxidation of proteins and its relationship with functional state of organisms. *Biochem. (Moscow). 72(8):* 809-827.

Lushchak, V. I. (2011). Environmentally induced oxidative stress in aquatic animals. *Aquat Toxicol. 101(1):* 13-30.

Mansouri, B., Hoshyari, E., and Mansouri, A. (2011). Study on ambient concentrations of air quality parameters (O_3, SO_2, CO and PM_{10}) in different months in Shiraz city, Iran. *Interna. J. of envir. Sci.*, *1(7)*, 1440.

Markert, B. A., Breure, A. M., and Zechmeister, H. G. (2003). Definitions, strategies and principles for bioindication/biomonitoring of the environment. *Trace Meta. and other Contam. in the Enviro.* 6: 3-39.

Maroni, G., and Watson, D. (1985). Uptake and binding of cadmium, copper and zinc by *Drosophila melanogaster* larvae. *Insect biochem. 15*(1): 55-63.

Martínez-Paz, P., Morales, M., Martínez-Guitarte, J. L., Morcillo, G., (2013). Genotoxic effects of environmental endocrine disruptors on the aquatic insect *Chironomus riparius* evaluated using the comet assay. *Mutat. Res. 758*: 41e47.

Mason, B. J. (1983). *Preparation of soil sampling protocol: techniques and strategies.* Environmental Monitoring Systems Laboratory, Office of Research and Development, *US Environmental Protection Agency.*

Mazhoudi, S., Chaoui, A., Ghorbal, M. H., and El Ferjani, E. (1997). Response of antioxidant enzymes to excess copper in tomato (*Lycopersicon esculentum*, Mill.). *Plant Sci.127*(2), 129-137.

Meng, Z., Qin, G., and Zhang, B. (2005). DNA damage in mice treated with sulfur dioxide by inhalation. Environ. *Molec. Mut. 46(3):* 150-155.

Migula, P., Laszczyca, P., Augustyniak, M., Wilczek, G., Rozpedek, K., Kafel, A., and Woloszyn, M. (2004). Antioxidative defense enzymes in beetles from a metal pollution gradient. *Biologia (Bratisl.). 59*: 645–654.

Misra, H. P., and Fridovich, I. (1972). The role of superoxide anion in the autoxidation of epinephrine and a simple assay for superoxide dismutase. *J Biol Chem. 247(10):* 3170-3175.

Moor, C., Lymberopoulou, T., and Dietrich, V. J. (2001). Determination of heavy metals in soils, sediments and geological materials by ICP-AES and ICP-MS. *Microchimica Acta.; 136(3-4):*123-128.

Morales, M., Martínez-Paz, P., Oz_aez, I., Martínez-Guitarte, J.L., and Morcillo, G., 2013. DNA damage and transcriptional changes induced by tributyltin (TBT) after short in vivo exposures of Chironomus riparius (Diptera) larvae. *Comp. Biochem. Physiol. C Toxicol. Pharmacol. 158*: 57e63.

Mukhopadhyay, I., Chowdhuri, D. K., Baypayee, M., Dhawan, A., (2004). Evaluation of *in vivo* genotoxicity of cypermethrin in *Drosophila melanogaster* using the alkaline Comet assay. *Mutagenesis. 19*: 85-90.

Naderizadeh, Z., Khademi, H., and Ayoubi, S. (2016). Biomonitoring of atmospheric heavy metals pollution using dust deposited on date palm leaves in southwestern Iran. *Atmósfera, 29(2),* 141-155.

Nath, S., Roy, B., Bose, S., and Podder, R. (2015). Impact of arsenic on the cholinesterase activity of grasshopper. *Am-Eurasian J Toxicol Sci 7:*173–176

Okamoto, T., Taguchi, M., Osaki, T., Fukumoto, S., and Fujita, T. (2014). Phosphate enhances reactive oxygen species production and suppresses osteoblastic differentiation. J Bone. *Miner. Metab. 32(4):* 393-399.

Prahalad, A. K., Inmon, J., Dailey, L. A., Madden, M.C., Ghio, A. J., and Gallagher, J. E. (2001). Air pollution particles mediated oxidative DNA base damage in a cell free system and in human airway epithelial cells in relation to particulate metal content and bioreactivity. *Chem. Res. Toxicol. 14(7)*: 879–887.

Radojevic, M., Bashkin, V. N. (1999). Practical environmental analysis. *Royal Soci. of Chem.*

Rhoden, C. R., Wellenius, G., Ghelfi, E., Lawrence, J., Gonzalez-Flecha, B. (2005). PM-Induced Cardiac Oxidative Stress Is Mediated by Autonomic Stimulation. *Biochem. Biophys. Acta. 172:* 305–313.

Rojas, E., Lopez, M. C., and Valverde, M. (1999). Single cell gel electrophoresis assay: methodology and applications. *J Chrom. (B): Biomed. Sci. App. 722(1):* 225-254.

Schmidt, G. H., and Ibrahim, N. M. (1994). Heavy metal content (Hg^{2+}, Cd^{2+}, Pb^{2+}) in various body parts: its impact on cholinesterase activity and binding glycoproteins in the grasshopper *Aiolopus thalassinus* adults. *Ecotoxicol Environ 29:*148–164.

Schmidt, G. H., Ibrahim, N.M., and Abdallah, M.D. (1992). Long-term effects of heavy metals in food on developmental stages of *Aiolopus thalassinus* (Saltatoria: Acrididae). *Arch Environ Contam Toxicol 23:*375–382

Sena, L. A., and Chandel, N. S. (2012). Physiological roles of mitochondrial reactive oxygen species. *Mol. Cell. 48:*158–167.

Seyyedi, M. A., Farahnak, A., Jalali, M., and Rokni, M. B. (2005). Study on Glutathione -S-Transferase (GST) Inhibition Assay by Triclabendazole. I: Protoscoleces (Hydatid Cyst; *Echinococcus granulosus*) and Sheep Liver Tissue. *Irani. J.of Public Health. 34*(1): 38-46.

Sharma, A., Shukla, A.K., Mishra, M., and Chowdhuri, D.K. (2011). Validation and application of *Drosophila melanogaster* as an *in vivo* model for the detection of double strand breaks by neutral Comet assay. *Mutat. Res. 721*: 142–146.

Shinkai, Y., Li, S., Kikuchi, T., and Kumagai, Y. (2015). Participation of metabolic activation of 2, 4, 6-trinitrotoluene to 4-hydroxylamino-2, 6-dinitrotoluene in hematotoxicity. J. *Toxicol. Sci. 40(5):* 597-604.

Shukla, A. K., Pragya, P., and Chowdhuri, D. K. (2011). A modified alkaline Comet assay for in vivo detection of oxidative DNA damage in Drosophila melanogaster. *Mut Rese/Gene Toxicol. Environ. Mut., 726(2)*: 222-226.

Shukla, S., Jhamtani, R. C., Dahiya, M. S., and Agarwal, R. (2017). Oxidative injury caused by individual and combined exposure of neonicotinoid, organophosphate and herbicide in zebrafish. *Toxic. Reports., 4*: 240-244.

Siddique, H. R., Gupta, S. C., Dhawan, A., Murthy, R. C., Saxena, D. K., and Chowdhuri, D. K. (2005). Genotoxicity of industrial solid waste leachates in Drosophila melanogaster. *Environ. Molec. Mut. 46(3),* 189-197.

Sofo, A., Scopa, A., Nuzzaci, M., and Vitti, A. (2015). Review. Ascorbate Peroxidase and Catalase activities and Their Genetic Regulation in Plants Subjected to Drought and Salinity Stresses. *Int. J. Mol. Sci., 16:* 13561-13578; doi:10.3390/ijms160613561

Sureda, A., Box, A., Enseñat, M., Alou, E., Tauler, P., Deudero, S., and Pons, A. (2006). Enzymatic antioxidant response of a labrid fish (*Coris julis*) liver to environmental caulerpenyne. *Comp Biochem Physiol(C). 144(2):* 191-196.

Tice, R. R., Agurell, E., Anderson, D., Burlinson, B., Hartmann, A., Kobayashi, H., and Sasaki, Y. F. (2000). Single cell gel/comet assay: guidelines for *in vitro* and *in vivo* genetic toxicology testing. *Environ and Molec Mut, 35(3)*: 206-221.

Tice, R., Vasquez, M., (1999). Protocol for the Application of the pH>13 Alkaline Single Cell Gel (SCG) Assay to the Detection of DNA Damage in Mammalian Cells. Date of access: October 8, 2008. Available at: http://cometassay.com/Tice%20and% 20Vasques.pdf.

US EPA (2013). Standard Operating Procedure of surface water sampling

US EPA (2014) Cleaning up the Nations Hazards Wastes Sites. United States Environmental Protection Agency.

Valverde, M., and Rojas, E. (2009). Environmental and occupational biomonitoring using the Comet assay. *Mut Res/Rev in Mut. Res. 681(1):* 93-109.

Vij, P. (2015). Environmental Pollution: Its Effects on Life and its Remedies. *AADYA-National Journal of Management and Technology (NJMT)*, *3*(2): 88-92.

Wilczek, G., Kramarz, P., and Babczyńska, A. (2003). Activity of carboxylesterase and glutathione S-transferase in different life-stages of carabid beetle (*Poecilus cupreus*) exposed to toxic metal concentrations. *Compa. Biochem. and Physi. Part C: Toxicology & Pharma.*, *134*(4): 501-512.

Winston, G. W. (1991). Oxidants and antioxidants in aquatic animals. *Comp. Biochem. Physiol. (C)*. *100*: 173-176.

Woodring, J.P., and Sparks, T.C., (1987). Juvenile hormone esterase activity in the plasma and body tissue during the larval and adult stages of the house cricket. *Insect Biochem. 17:* 751–758.

Wuana, R. A., and Okieimen, F. E. (2011). Heavy metals in contaminated soils: a review of sources, chemistry, risks and best available strategies for remediation. *ISRN Ecol.* 20 pages.doi.org/10.5402/2011/402647

Yan, B., Wang, L., Li, Y., Liu, N., and Wang, Q. (2007). Effects of cadmium on hepatopancreatic antioxidant enzyme activity in freshwater crab *Sinopotamon yangtsekiense*. *Acta Zool. Sin. 53 (6):* 1121–1128.

Yousef, H. A., Abdelfattah E. A. and Augustyniak M. (2017). Evaluation of oxidative stress biomarkers in *Aiolopus thalassinus* (Orthoptera: Acrididae) collected from areas polluted by the fertilizer industry. *Ecotoxic.* doi:10.1007/s10646-017-1767-6.

Yousef, H. A., Afify, A., Hasan, H. M., Meguid, A. A., (2010). DNA damage in hemocytes of *Schistocerca gregaria* (Orthoptera: Acrididae) exposed to contaminated food with cadmium and lead. *Nat. Sci. 2*: 292–297.

Zhang, Y., Sun, G., Yang, M., Wu, H., Zhang, J., Song, S., and Guo, Y. (2011). Chronic accumulation of cadmium and its effects on antioxidant enzymes and malondialdehyde in *Oxya chinensis* (Orthoptera: Acridoidea). *Ecotox. and environ. Safety. 74*(5): 1355-1362.

Zhou, Q., Zhang, J., Fu, J., Shi, J., and Jiang, G. (2008). Biomonitoring: an appealing tool for assessment of metal pollution in the aquatic ecosystem. *Analytica chimica acta, 606*(2): 135-150.

Zhu, H., Zhang, J., Kim, M. T., Boison, A., Sedykh, A., Moran, K. (2014). Big data in chemical toxicity research: The use of high-throughput screening assays to identify potential toxicants. *Chem Res Toxicol. 27(10):* 1643-1651.

Printed by Books on Demand GmbH, Norderstedt / Germany